WITHDRAWN FROM
COLLECTION

CONTENTS

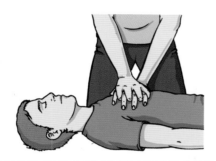

FIRST AID

SOS

SIGNALING

TRACKING

INTRODUCTION

There's nothing quite like an adventure in the great outdoors! Our amazing planet has millions of different places to discover, and exploring them is an enormous privilege—one that I have been lucky enough to take advantage of many times. It's very important to be fully prepared for an adventure, and know all the vital skills needed to survive in the wild before you set off. This book will teach you how to plan for an expedition, what you need to know before you go exploring, and what to take with you for an adventure in any environment. You will also learn some vital first aid skills, pick up tips on how to signal to the rest of your team and call for help in an emergency, and try your hand at tracking all sorts of different birds and animals in the wild. We have a responsibility to look after our world and keep it safe, so be sure to leave everything exactly as you found it. Remember—take only photographs, and leave only footprints!

EXPEDITION PLANNING

The success of any adventure relies on careful planning beforehand. This section will teach you what to take with you, how to create a camp, and how to avoid dangers when out exploring the wild. Above all, it is here to help you fully enjoy the adventure!

Bear

IN THIS SECTION:

AIMS AND OBJECTIVES

Before starting out, it's a good idea to plan out your aims and objectives. Sit down with your team beforehand and decide what you want to achieve—whether it's a new challenge, exploring a new place, or just a fun adventure in the wild.

Why go on an expedition?

An expedition is an extended journey through the wild. It probably won't be easy and may well have some tough moments. So why do it? Here are some of the benefits of going on an expedition:

- Fun and adventure in new surroundings.
- Explore the natural world.
- Learn new skills such as fire lighting and navigation.
- Be part of a team, and develop leadership skills.
- Challenge yourself and boost your confidence.
- Help a community in need.
- Raise funds for a cause close to your heart.

Location

First, you will need to decide on the location. Some expeditions visit remote, faraway places, while others explore closer to home. Whatever your plans, you need to research your destination carefully, to find out what conditions might be and what you will need to bring.

Planning

The success of any expedition depends on careful planning. This includes travel to and from your destination. Plan your trip in three stages: the journey there, the expedition itself, and the return home.

Type of expedition

How will you explore? There are many different methods of travel. Many expeditions involve several types.

BEAR SAYS

Take time to plan all aspects of your expedition. How big is your team? How long will the trip last? Maybe don't be overly ambitious on your first attempt!

horseback riding

canoeing or kayaking

cycling

CLOTHING

Having the right gear is crucial to the success of any expedition. Finding out about climate and conditions in advance is very important, as this will help you decide what to pack. The clothes, footwear, and equipment you will need might vary a lot depending on the weather and terrain of the area you're exploring.

Expedition gear

This picture shows basic clothing for an expedition. It's best to prepare for all sorts of weather. Expeditions to extreme environments may involve special equipment.

Layer method

Wear several light layers to keep comfortable in changing conditions. Each layer traps warm air. Peel off a layer if you get too hot, and add a layer if you feel cold.

snow goggles

beanie or sun hat

scarf or bandana

layers of clothing

water- and windproof jacket

gloves

waterproof pants

hiking boots

base layer traps warm air next to your skin

mid layer

outer layer and waterproof shell

Things to remember

Clothing should be loose fitting and comfortable. Avoid jeans, as they soak up water if it rains.

BEAR SAYS

Pack spare clothes in a waterproof bag to keep them dry. If your boots get wet, pack them with newspaper and put them in a warm place to dry out.

gloves or mittens will help keep your hands warm

a first aid kit including sunscreen is vital

hats are a must for all expeditions— pack one for sun, rain, or cold weather, depending on the weather forecast

Footwear

A pair of stout shoes or boots are vital when hiking. Make sure your boots are well "broken in" but not too worn. Clean and waterproof your footwear. Hiking socks help to prevent blisters. You may need lightweight shoes for around camp as well.

carrying a tin of boot wax will allow you to keep your boots waterproofed and in good condition

wear your boots to "break them in" before an expedition to avoid blisters and make sure they are comfortable to hike in

hiking socks will help protect your feet and keep them warm

SHELTER

Most trips that last more than one day will involve at least some camping. A good-quality tent, sleeping bag, and mat will keep you dry and comfortable overnight.

Tents

Your tent is your base and home. It also protects you from the elements. Tents come in many shapes and sizes. Your choice will depend on group size and the conditions you are likely to experience.

BEAR SAYS

Practice putting up and taking down your tent, including in windy conditions. Use a mallet to hammer in tent pegs securely and make sure the ropes are taut.

dome tents are easy to erect but can be tricky to take down and pack away

tube tents provide extra room for storage

ridge tents are built around an A-frame and are very sturdy but can be bulky to carry

Sleeping mats

Sleeping on the ground can be pretty uncomfortable. Luckily, there are several different types of sleeping mats or camp beds you can use to provide comfort and insulate you from the cold and wet ground, ensuring a good night's sleep!

camp beds are comfy
but heavy to carry

camping mats are basic, but are lightweight and easy to carry (in the military we used to cut the roll mat down so that it took up less space in our packs—we would cut it down to just cover our top half which is the key part of the body, the core, to keep warm)

a camping mattress can be packed up very small and inflated using a small foot or hand pump

Sleeping bags

A good sleeping bag will keep you cozy overnight, and is one of the most important pieces of equipment for any expedition. Research different types of sleeping bag before you go, so you have the right one for your trip.

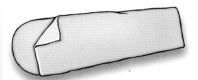

sheet sleeping bags
are good for hot
climates and if you're
sleeping in hostels

there are plenty of different types of sleeping bags to choose from—do your research beforehand to pick the correct bag for the correct environment

stuffing a sweater inside
your sleeping bag case
makes a good pillow

FOOD, WATER, AND COOKING

Food and water provide the fuel you need to stay active, fit, and healthy. Hot food and/or drink are a must after a long, cold day's hike, and will also help to boost your spirits. You may be able to buy or find food as you go along, but you should always bring supplies with you.

Expedition supplies

You need a lot of stores for a multi-day expedition. Plan a detailed menu so you take the right ingredients. All team members should lend a hand with cooking. Dried foods and canned foods will keep for a long time, though canned foods can be quite heavy.

BEAR SAYS

Store foods in a dry place, out of reach of animals such as mice, rats, raccoons, and bears. Hanging food high up in trees is the best bet. Use paracord.

Food pyramid

A balanced diet is vital to keep expedition members healthy. You need a varied diet containing carbohydrates, protein, fruits and vegetables, and vitamins and minerals every day. You also need fats, but avoid too much fatty and sugary food.

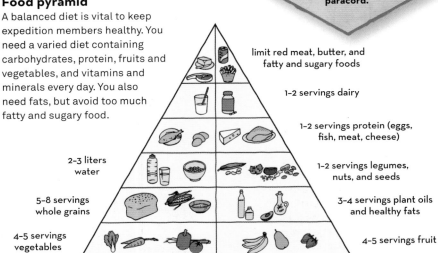

limit red meat, butter, and fatty and sugary foods

1–2 servings dairy

1–2 servings protein (eggs, fish, meat, cheese)

1–2 servings legumes, nuts, and seeds

3–4 servings plant oils and healthy fats

4–5 servings fruit

2–3 liters water

5–8 servings whole grains

4–5 servings vegetables

Trail snacks

Trail snacks provide energy while you're on the move. Sugary snacks can provide short bursts of energy, but it is better to go for foods like fruit, nuts, and protein-based snack bars.

Local foods

Most expeditions use local foods to add to their stores. You can also forage for wild foods, but you need to be absolutely sure to identify the right species, as some plants and many fungi contain deadly poison.

Drinking water

You need to drink at least 2–3 liters of water a day—more if you are active or in a hot climate. You will need to purify all water sourced from the wild, so bringing water purification tablets or iodine drops with you is a good idea.

Stoves and cooking gear

Stoves use different fuels, such as gas, paraffin, or solid fuel. You also may be able to cook over your campfire. Your cooking kit should include matches, pots and pans, cutting board and knife, wooden spoon, and spatula, plus mug, bowl, plate, and cutlery. Always wash them thoroughly after use.

HEALTH AND FIRST AID

It is vital that all expedition members have at least a basic knowledge of first aid. Take a first aid class and practice basic techniques before setting off on a major trip. Make sure you know any medical needs of expedition members in advance, and mark all medications clearly. For more information, see pages 52–95.

First aid kit

Pack a full first aid kit, including bandages, dressings, tape, scissors, tweezers, and disposable gloves.

First aid drill

In the event of an accident, keep calm and assess the situation. Get everyone out of danger, assess the injury, and contact the emergency services straightaway if possible.

Emergency ABC

If an accident strikes remember ABC— Airway, Breathing, Circulation.

First, talk to the patient to see if they are conscious. Shout if needed. If they don't respond:

A. Tilt their head back to clear their **airway**. Make sure nothing is lodged in their throat.
B. Check to see if the patient is **breathing**. Is their chest rising? Hold a mirror under their nose to see if their breath fogs it up.
C. Check their **circulation** by checking their wrist or neck for a pulse.

BEAR SAYS

Wash your hands thoroughly before treating an injury. Wear disposable gloves, especially when treating open wounds such as cuts and burns.

Recovery position

Place an unconscious patient in the recovery position to keep their airway clear and avoid further injury. Roll them onto their side. Bend their upper arm under their lower cheek and place their upper leg at a right angle to their body.

Cuts

Wash the affected area thoroughly to avoid dirt getting into the wound and causing an infection. Press a cloth or dressing on the wound to staunch the bleeding, then cover the wound with a bandage or clean dressing.

Bites and stings

If the stinger is left in the wound, carefully remove it with tweezers. Wash the wound site and cover with an iced or cold dressing. If necessary, ask an adult to provide the patient with painkillers.

Burns

Run the wound site under cold or lukewarm water. Do not use ice, as this may damage the skin further. Cover the burn with a nonstick dressing, plastic wrap, or even a clear plastic bag. Ask an adult to provide painkillers if needed.

OTHER GEAR

Expeditions require all sorts of gear besides food, camping, and first aid equipment. Think about what you will need during the day, when you camp, and overnight.

notebook and pencils

portable GPS

flashlight

mobile phone in a waterproof case and a portable charger

waterproof matches

compass and map

flint and steel

pocketknife

watch

BEAR SAYS

Read through the instructions on all equipment. Practice to make sure you know how to use all your gear.

whistle

backpack

Emergency shelter

A rope or paracord and tarpaulin will help you to rig an emergency shelter anytime, anywhere.

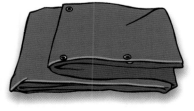

Repair kit

A needle and thread, clear tape, safety pins, rubber bands, and superglue will help you mend tears and rips to clothing, maps, tents, and other gear.

Personal kit

A toiletry bag should contain personal items such as soap, shampoo, toothbrush, and toothpaste. Hygiene is very important on an expedition to avoid getting sick or risking infection. Don't forget sunscreen and lip balm to protect your skin from the sun.

Extras

Consider taking these extras, but don't let your pack get too heavy!

playing cards

binoculars

camera

WOODS AND FORESTS

Do your homework before setting out on an expedition. Study maps of the region. Find out about climate, wild foods, natural hazards, and dangerous animals. Woods and forests offer sheltered conditions, but route finding is difficult.

Navigation

Navigation is tricky in woods and forests because you can't see the way ahead. Dense vegetation can hide hazards such as sheer cliffs and deep gorges. Check the map for landmarks such as rivers, bridges, buildings, and settlements. Take advantage of clearings and high ground to scout the area ahead.

Finding your way

It's hard to move in a straight line through woodland. In fact, it's all too easy to go around in a big circle! Use a compass bearing (see pages 46–47) to head in the right direction. Send one person ahead in that direction, then have them stop before they go out of sight. Join them and repeat.

Communication

All members of the party should carry a whistle so you can communicate if out of sight. Work out some simple signals everyone knows, for example, five whistle blasts means return to base.

Dangerous animals

Some forests contain predators such as wolves and bears. Suspend food on a rope looped over a high branch to avoid attracting bears and other scavengers.

Building a debris shelter

1. Wedge a long, straight branch against a tree trunk or in a tree fork to make a ridgepole.
2. Lay smaller branches against it to create a tent shape.
3. Weave twigs, leaves, and debris between the branches to provide shelter from wind and rain.

BEAR SAYS

Wild foods such as nuts and berries may be found in woods and forests, but great care is needed to identify them, as they may be deadly.

DESERTS AND DRY PLACES

Less than 10 inches of rain falls each year in a desert, creating a very dry, often very hot environment. Very few plants and animals can survive in deserts, meaning that food, water, and shelter can be very difficult to come by.

Save water

Water is the body's main need. You cannot survive for more than a few days without it. Make sure you have adequate water supplies before venturing into a desert. Conserve water by resting in the shade by day and working or traveling at night.

Finding water

Study your map for likely water sources. Oases form where underground water seeps to the surface. You can also fill water tanks at wells. A dried-up riverbed may hold water if you dig below the surface. Plants growing on a cliff will be watered by a spring or seep, so keep an eye out for plant life. It is also possible to find water inside some species of cactus, like the barrel cactus.

BEAR SAYS

If your vehicle breaks down, stay in its shade and wait for another vehicle to pass. Don't head off into the desert.

Climate and clothing

Deserts have extreme temperatures—scorching hot by day, freezing cold at night. Wear lightweight, loose-fitting clothes. Most people who live in the desert wear black, as it expels heat from your body. Light colors will absorb less heat, but will not reflect it as well.

a broad-brimmed hat or head cloth protects your head and the back of your neck from the burning sun

long-sleeved shirts and pants protect against sunburn

Poisonous animals

Dangerous desert animals include venomous snakes and scorpions. A scorpion's tail is tipped with a painful stinger.

boots guard against snakes and scorpions—don't wear flimsy sandals!

Building a desert scrape

1. Dig a pit long and deep enough to lie down in.
2. Cover with a tarpaulin. Weigh the edges down with rocks.
3. A double layer of cloth will keep the air cooler below.

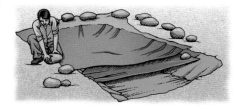

JUNGLE AND TROPICS

Jungles are among the toughest environments on Earth. Torrential rain falls almost daily, and these dense, dark forests teem with creepy crawlies. Travel is incredibly difficult, except by river.

Navigation

Route finding is tricky in the rain forest. Keep to the trail if there is one—people have gotten lost after taking just a few steps off the track. If there is no trail, you might have to hack one with a machete, using your compass to keep on course.

BEAR SAYS

Progress will be very slow if you are hacking your way with a machete. Plan to cover just a short distance each day.

Climate

Over 80 inches of rain falls each year in a rain forest. Thunderstorms strike on most days, and flooding is a danger in the rainy season.

Gear and clothing
- Wear a long-sleeved shirt and pants to guard against biting insects.
- A wide-brimmed hat will keep the sun off.
- High boots protect against leeches when crossing streams.
- Essential equipment includes a machete, compass, GPS, and mosquito net.

Wildlife
Large mammals such as elephants and rhinos live in rain forests. Tigers and jaguars are top predators. Many snakes, frogs, spiders, and centipedes are armed with poison, while leeches, ticks, and vampire bats will suck your blood.

Jungle bivouac
1. String a rope between two trees at head height and tighten until taut.
2. Throw a tarp over the rope and peg the ends securely.
3. Now rig a hammock or camp bed at waist height, out of reach of creepy crawlies.

✗ BEAR SAYS
Bugs and germs breed quickly in the tropics. Make sure all wounds are sterilized and take extra care to purify your water.

COLD PLACES

Expeditions to the tundra and polar regions face a tough survival challenge. Storms, ice-covered lakes, and predators are serious hazards, while hypothermia and frostbite can strike quickly in icy conditions.

Climate

The polar regions are the coldest places on Earth. Temperatures only rise a few degrees above freezing in summer and can drop to −40°F in winter. Strong winds make the air seem even colder. It's light for 24 hours a day in summer, and dark all day in winter.

Cold-weather hazards

Hypothermia is when your body temperature drops dangerously low. Hot food or drinks will help you recover. Frostbite is when the skin starts to freeze. Check the nose, ears, fingers, and toes for signs of frostbite.

Gear and clothing

Wear several layers of warm clothing, including thermal underwear. The outer layer should be wind- and waterproof pants and a jacket with a warm hood. Goggles prevent snow blindness. You may also need snowshoes, a snow shovel, GPS, and emergency flares.

frostbite

Blizzards

In a blizzard, whirling snow fills the air, and visibility falls to zero. Don't try to travel when the weather's like this. Stay in your tent, but go outside regularly to clear heavy snow from the fabric, or the tent could collapse under the weight.

Predators

Polar bears live in the High Arctic and are among the most dangerous animals on Earth, so should be avoided. If you are in an area where polar bears are common, it might be worth hiring an armed guard. Grizzly bears also hunt on the tundra, while walruses and huge elephant seals live on polar coasts.

Snow trench shelter

1. Dig a long snow trench about 1.5 feet deep.
2. Excavate the snow in blocks about 15 x 20 inches and 6 inches thick.
3. Lean the blocks together to form a pitched roof over the trench.

BEAR SAYS

Snow-covered landscapes have very few landmarks, so route finding is difficult. Use a compass to avoid going around in circles.

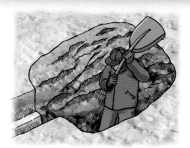

HILLS AND MOUNTAINS

Mountains have a harsh, cold climate with rapidly changing conditions. The weather can change from baking hot to freezing wind with very little warning. Hazards include glaciers, avalanches, landslides, scree, mist, and fog.

Climate

The temperature drops 1°F for every 500 feet you climb. The air high on mountains also contains less oxygen. Above 26,250 feet, there is so little oxygen that the body can't survive for long—this region is known as the Death Zone. If you spend time in these conditions, you may suffer from an illness called "altitude sickness," which can be deadly. This can strike at much lower elevations as well. If in doubt, go down.

Equipment and clothing

Wear several layers of warm clothing to guard against hypothermia. Goggles and a wool or fleece hat are essential. Climbing gear includes ice ax, harness, rope, and crampons— metal spikes fixed to your boots. Skis, ski poles, and an avalanche probe might come in handy too.

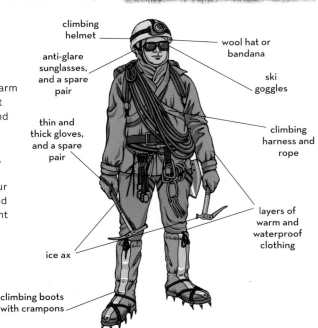

climbing helmet

anti-glare sunglasses, and a spare pair

wool hat or bandana

ski goggles

thin and thick gloves, and a spare pair

climbing harness and rope

layers of warm and waterproof clothing

ice ax

climbing boots with crampons

Mountain camps

Finding a place to camp on a mountain can be difficult. Climbers will bivouac on steep slopes, dips, and saddles, often having to hack out a small area of flat ground to make camp. Beware avalanche slopes above, and always avoid camping in such dangerous zones.

Glaciers

These are masses of ice that carve out valleys through the mountains and slowly flow downhill. They are cut by deep cracks called crevasses. You have to take special care not to fall into these, especially if the cracks are covered by fresh snow.

Avalanches

An avalanche is when a mass of snow breaks away and thunders downhill, often with terrifying speed. Avalanches can be triggered by the sun's heat or by a climber or skier. The key with avalanche survival is simply to avoid being in a high avalanche situation. If you are caught in an avalanche, use a swimming motion, and fight to stay near the surface. Cover your nose and mouth to avoid inhaling powder snow. When it stops, act fast, and try to kick or claw your way out. Spit to see which way it falls to make sure you are digging in the correct direction, as an avalanche can be very disorientating.

Building a snow-hole bivouac

Dig a snow hole in a deep bank of snow. The entrance tunnel should slope down, then upward, so the cold air will sink. Use a stick or ski pole to make an air hole, and build a sleeping platform.

BEAR SAYS

If thick mist descends, you will need to use your compass to find your way. If you have a GPS device, use it to locate your exact position.

ON OR BY WATER

Many expeditions travel by water and camp by rivers, lakes, or on seashores. This is often a useful method of travel, especially in thick jungle or places where the terrain is harsh. However, these watery places hold their own dangers, such as strong currents, tides, rapids, and floods.

Gear and clothing

Canoeists and sailors need special equipment. You will need wind- and waterproof gear or a wetsuit, plus a lifejacket and safety helmet.

Capsize drill

All kayakers learn the capsize drill, so they know what to do if their craft overturns. If your kayak has a spray deck, pull off the cover and kick your way to the surface. Expert kayakers can roll right over, but it takes a lot of practice to be able to do this.

Abandon ship

If you have to abandon ship, keep your clothes on but remove your shoes, as they will weigh you down. Enter the water gradually if possible. Swim away from a sinking ship.

Treading water

Once in the water, inflate your life jacket or grab a floating object. Don't exhaust yourself swimming unless you are near the shore—just tread water to stay afloat. Crossing your hands and feet and keeping your head dry will help you keep warm.

BEAR SAYS

Beware riptides that flow away from the shore. If caught in one, don't swim against the current but calmly swim across it, until you are out of it.

BEAR SAYS

Beware waves, tides, and currents in the sea and rivers. Never swim alone. On the seashore, make sure you don't get cut off by the tide.

Wild foods

Fresh and saltwater habitats offer foods such as seaweed, fish, and shellfish. You can catch fish with a simple rod and line, but you will need to be very patient. Prepare a fish for cooking by slitting its belly to remove the guts and bones.

Predators

Deadly sea creatures include sharks and poisonous jellyfish, while crocodiles, large snakes, and hippos may lurk in rivers and swamps. Make sure you know what predators are native to the area you are exploring and how to stay safe around them. Don't take chances with crocodiles or hippos—especially in murky water.

River crossing

First, scan the riverbank for the best crossing place. Cross using a stout stick for support, or link arms and cross in groups of two or three. Use any flotation you can find. If in doubt, do not enter rivers. They are incredibly dangerous even when apparently slow moving.

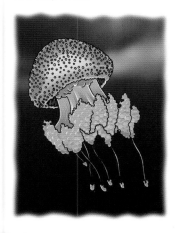

Build a raft

You can make a raft using lengths of bamboo lashed together with cord. Tie crosspieces to make the craft more stable. Two layers of bamboo will make the raft float better. Make paddles out of wood or bamboo.

FITNESS AND TRAINING

You need to be fit for an expedition. Increase your general fitness with regular exercise. You will also need to train for any specific activities your trip involves, such as hiking, cycling, climbing, or canoeing.

Fitness routine

Get in good shape through well-planned fitness training. Aim to do half an hour of exercise at least four days a week. If you are unfit, start with lower intensity exercise and build up. This could include time spent doing sports, or playing games, gym, dance, brisk walking, or chores such as housework or gardening.

Aerobic exercise

Aerobic means "with oxygen," and includes any exercise that will make your heart and lungs work harder. Aerobic activities, such as jogging, cycling, swimming, or rowing, will improve your stamina over sustained times.

Keeping records

Keep a record of your training. Time yourself over a distance using a stopwatch, or count the number of times you can do an exercise within a specific time such as a minute. Aim to improve your time or increase the intensity and number of exercises you do every week.

Strength and flexibility

Strength exercises such as sit-ups, pull-ups, and push-ups help to build general muscle strength, while stretching exercises make you supple. This will help with any physical activity, but especially activities that require specific muscle groups like cycling, climbing, or canoeing.

Practice makes perfect

You should always practice the specific activity involved in your trip beforehand, in order to build strength and stamina. If your trip will involve covering a certain distance every day, start with a shorter distance in your training, then build up to the distance you will need to do.

Warm up and cool down

It is important to warm up and cool down before and after training by doing stretching exercises, otherwise you may risk damaging your muscles.

BEAR SAYS

Plan gradually and steadily to increase the intensity of your training. Don't go too hard too soon, or you could strain a muscle.

TEAMWORK

Some explorers like to go it alone, but most expeditions involve a whole team of people with a range of different strengths and skills. Not only is going in a group much safer, it offers you the chance to share your adventure and have more fun!

Leadership

Most expeditions are led by one person, though you can also decide to take turns as leader. Being a team leader is rewarding, but also brings responsibility. You are in charge of safety and the smooth running of the expedition, and may need to keep up people's morale. A good leader will make use of people's different talents and will always help those who are struggling.

Pros and cons

Group expeditions have many advantages:

- Safety in numbers.
- Several brains to share decisions such as route finding.
- Sharing equipment such as tents and supplies reduces costs.
- Companionship and fun.

Everyone in the group has to be willing to play their part and share the chores.

Decision making

A group leader should be open to everyone's ideas and suggestions, but she or he may also need to make tough decisions that not everyone agrees with. Always explain your decisions to your team and take people's feelings into account.

Being a team player

Being part of a team is a fantastic experience, but may sometimes involve accepting decisions you aren't entirely happy about. It is OK to discuss these things, but sometimes be prepared to compromise. Work together to make the team something you can all be proud of!

Building teamwork

Before you set off, practice the activities involved in the expedition together. This will help you get to know one another and find out about strengths and weaknesses. If things don't go to plan, analyze what went wrong. How could the team improve its performance?

Support team

All expeditions should have a backup team which will provide support and may help to solve problems. When setting off on local expeditions, always tell a parent or responsible adult when and where you are going and what time you expect to be back. This means that, if something goes wrong and you do not arrive home, they will be able to send for help.

BE PREPARED

The Scout motto, "Be Prepared," is excellent advice for all expeditions, whether close to home or far away. Plan each stage of the trip carefully in advance, and always make a backup plan in case things go wrong.

Dealing with "red tape"

Foreign and sometimes even local travel involves meeting certain official requirements. If you are traveling abroad, you will need a passport and possibly also a visa. You need tickets and money in the local currency. It is also important to find out in advance if you meet all the medical requirements for the region you are visiting, and have all the vaccinations and medication needed for the trip. For example, many regions will require you to take malaria medication.

Local guides

Many expeditions to remote places hire a local guide to show the way and provide expert advice on wild foods, poisonous plants, and dangerous animals. If you don't hire a guide, you will need to learn at least a few words of the language, research local customs, and find out how people live.

Timetable

Making a timetable will help you plan your adventure. Below is an example of what an expedition timetable may look like.

Day 1. Take the 4:00 p.m. 7a bus from Weston to Smalltown and walk to Campsite 1. ETA (estimated time of arrival): 5:15 p.m. Make camp and prepare meal.

Day 2. 6-mile hike from Campsite 1 to Campsite 2 at Northport. ETA: 4:00 p.m. Camp and prepare meal.

Day 3. 7-mile hike from Campsite 2 to Campsite 3 at Westport. ETA: 4:00 p.m. Camp and prepare meal.

Day 4. Catch 11:00 a.m. bus 10a from Westport to Weston, ETA 12:00 p.m. Walk home.

Emergency plan of action

Even the best-run expeditions don't always go to plan. Try to think about what could go wrong. Write an Emergency Plan of Action (EPA), giving details of what to do if things don't go to plan, for example if the weather is bad or someone gets ill.

EPA information sheet

Your EPA should contain an information sheet for all members of the expedition. This should include their full name and address, passport number, contact details, details of your trip, and contact details for the support team and close family—plus any medical issues. Each member should carry a copy of this. Make sure the EPA is always waterproofed.

PACKING

All expedition members need a rucksack or backpack to safely stow and carry their gear. You should pack carefully so that you can easily lay your hands on what you need, when you need it. Planning your packing beforehand will save you a lot of time and effort when you are out on your adventure.

Choosing a rucksack
There are several different types of rucksack. If buying one, try it on to see if it feels comfortable.

- Does it have a hip belt and all the features you need?
- Frame or inner-frame rucksacks allow air to circulate between your back and the rucksack, reducing sweat.
- A small daypack can be useful to carry items you need during the day.
- A strong, adjustable hip belt takes the weight off your shoulders.
- A chest strap helps to balance the pack.

Waterproofing
A good rucksack cover will help keep the rain off your rucksack. It's also important to use a waterproof "liner" or plastic bag inside the pack too, to keep the contents dry. You don't want to get to camp after a long day's hike in the rain only to find all your clothes and equipment soaked!

Packing

Storing selected gear in cloth or plastic bags before you pack them will help you find items quickly. Pack your gear in reverse order, so the things you will need first are on top. Put heavy items close to your back to stop them putting extra strain on your shoulders. Vital items such as your first aid kit should be at the very top, so they can be accessed quickly in an emergency.

BEAR SAYS

Carrying too much weight will slow you down and make the journey harder. Be smart and disciplined with both essentials and luxuries.

keep fragile items and things you will need during the day (waterproofed essentials, first aid kit) stored at the top

heavy items (such as your tent) should be stored next to your back

items that are only needed at the camp should be stored at the bottom

CAMPCRAFT

All the skills used in outdoor camping are known as campcraft. This includes putting up a tent, lighting a fire, and cooking outdoors. Organize your camp so it is safe, practical, and comfortable for everybody on your expedition.

Camp location

Choose a level spot for your camp, if possible near a source of wood and water. But don't camp in a valley bottom or by a stream or river that could flood after heavy rain. Avoid the banks of lakes where mosquitoes breed.

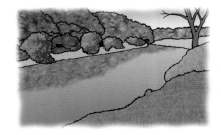

Organizing your camp

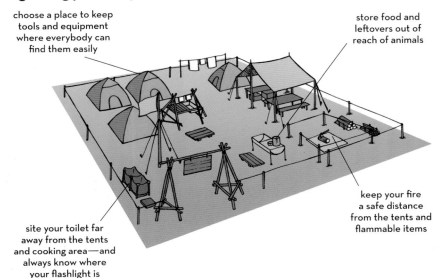

choose a place to keep tools and equipment where everybody can find them easily

store food and leftovers out of reach of animals

keep your fire a safe distance from the tents and flammable items

site your toilet far away from the tents and cooking area—and always know where your flashlight is

Fire making

To light a campfire, you will need tinder to take the spark, and kindling to fan it. You will also need different sizes of wood to feed the fire. Choose a sheltered location and clear the ground of debris. The easiest way to light a fire is by using matches or a lighter. A fire steel will also strike a spark.

BEAR SAYS

Beware dead branches on mature trees, which could fall on you or your tent. Falling dead branches are a big danger in jungles.

Fire safety

Fires are very dangerous. Make sure you make your fire a safe distance from your tents, and have water standing by so you can put the fire out if necessary. Always make sure your fire is completely out before you move on—you don't want to risk causing damage to the area.

Cooking tripod

You can hang a kettle over the fire using a tripod. You will need three strong, straight sticks of about the same length. Bind rope around one end, then splay the legs out, creating a tripod. Place the tripod over the fire, making sure it is stable, then attach the kettle on a hook.

NAVIGATION AND MAP READING

Everyone should learn the basics of navigation before setting out on an expedition. One of the most important navigational skills is knowing how to read a map. Explorers visiting unknown regions may draw their own maps for future expeditions.

Maps and symbols

Maps are drawings of the landscape from above, and help us find our way. They use symbols to show landmarks and permanent features such as woods, roads, and rivers. There is normally a "key" or "legend" at the side of the map which explains what the symbols mean.

Scale

Everything on the map is drawn to the same size—this is called the scale. There is a scale bar at the side of the map to show you what scale it is drawn to. For example, 1 inch on the map may represent 0.5 miles, 1 mile, or 10 miles. Studying the scale helps you judge distances and work out how long it will take to walk anywhere.

disabled access

art gallery

cafe

campsite

castle or fort

cycle trail

fishing

garden

golf course

information center

picnic site

church or cathedral

public bathrooms

restaurant

walks or trails

water

birdwatching

1 inch = 1 mile (1:63,360)

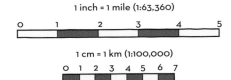

1 cm = 1 km (1:100,000)

Grid references

Squares on the map form a grid which can help you pinpoint locations. These are called grid references. To read or give a grid reference, start at the bottom left-hand corner of the map. Run your finger sideways along the map, then up. Remember this order using this phrase: "through the hall then up the stairs." Grid references give the East-West direction first, then the North-South direction.

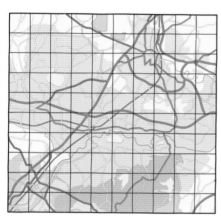

Contour lines

How do you show the ups and downs of the landscape on a flat map? The answer is using contour lines. These lines link places at the same height above sea level. Reading the contours shows you the location of hills and valleys, so you know if you have to climb or descend. If the contours are close together, the slope is steep.

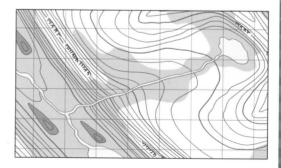

FINDING YOUR WAY

To work out what direction to travel in, you need to know how to use a compass. With a map and compass, you can find your way anytime, anywhere! A skilled map reader can also judge the terrain, distance, and roughly how long a journey will take. The key factors to navigate by are: bearing, distance, time, features, and backdrop (remember this using the words Bear Drinks Tea For Breakfast!).

Parts of a compass

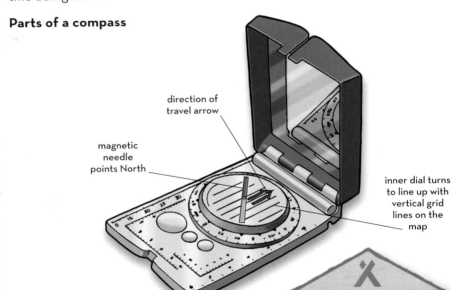

direction of travel arrow

magnetic needle points North

inner dial turns to line up with vertical grid lines on the map

Using a compass

A compass can be used in several different ways to find direction. The red magnetic needle always points North. This allows you to work out the other main compass points: South, East, and West, and all points in between.

BEAR SAYS

Grid squares on a map can help you judge time and distance. If you walk at 3 mph and each square is 1 mile, you will walk across roughly three grid squares in an hour at a steady pace.

Orient the map
This process allows you to find your direction with a map and compass.

1. Place the edge of the compass on the map along the route you intend to travel.
2. Turn the inner compass dial so the lines match up with the grid lines on the map. Make sure the red arrow on the dial points to North on the map (usually at the top).
3. Now take the compass off the map and hold it flat. Turn around until the red magnetic needle lines up with the red arrow on the dial. The direction of travel arrow now points to where you want to go.

Judging distance
A device called a map measurer works out distances on a map. Simply run the little wheel along your route and then read the distance. You can also lay a piece of string along your route. Then lay it along the scale bar, several times if necessary, to see how far you have to go.

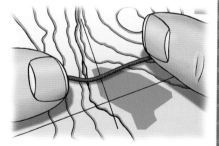

the Southern Cross

the Big Dipper

Navigating by night
If you don't have a compass, you can use the stars to navigate at night. In the Northern Hemisphere the Big Dipper (Great Bear) shows where North is. The two stars on the tip of the Big Dipper point toward Polaris, the North Star. In the Southern Hemisphere, the foot of the Southern Cross points South.

SIGNALING

Learn to send signals of various kinds before embarking on an expedition. Signaling not only helps the group to stay in contact, but it can also be used to call for help, which could save your life in an emergency. For more information, see pages 96–139.

Two types of signals

There are two main types of signals: visual and audio signals. These correspond with our two main senses, sight and hearing. Visual signals include flags, light flashes, direction arrows, and texts on your mobile phone. Audio signals include phone calls, shouts, and whistle blasts.

Natural materials

Natural materials can be used to mark a trail. You can also send a distress call by arranging sticks or rocks to spell SOS in large letters, or by drawing large letters in mud, snow, or sand.

BEAR SAYS

Distress calls are taken very seriously by the emergency services. Never send out an SOS unless it is a real emergency.

Morse code

This international code uses dots and dashes, or short and long signals, to spell out letters and words. The most important message in Morse code is SOS: three dots, three dashes, and three more dots.

Trail signs

Scouts use these signs to mark a trail for others to follow.

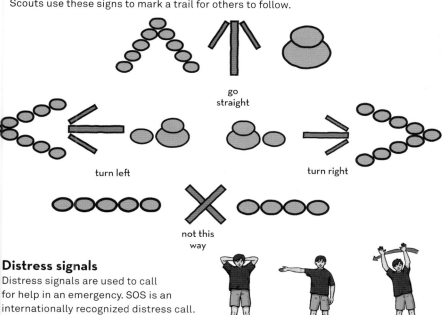

go
straight

turn left

turn right

not this
way

Distress signals

Distress signals are used to call for help in an emergency. SOS is an internationally recognized distress call. This signal can be sent in various ways, including in Morse code. Any signal sent three times, such as three light flashes or three whistle blasts, also acts as an emergency distress call. You can also use body signals to communicate with a helicopter or airplane.

our receiver
is operating

wait, I can
proceed shortly

do not
land here

use drop
message

all OK, do
not wait

land here

need mechanical
help

pick us up

no

yes

STAYING SAFE

Every expedition will meet with some difficulties or problems. When this happens, the most important thing is to stay calm and work out the best solution. If things don't work out as you hoped, stay positive and switch to your backup plan. It helps to be as flexible as possible. As the Commandos say: be comfortable with uncertainty. This is a good trait for the survivor.

Contact base

Make regular contact with your support team at set times, so they know that all is well. This way, if you don't contact them, they will know to send for help. Don't forget to take spare batteries or cables for charging equipment such as a radio or mobile phones—you don't want to cause an unnecessary panic over a flat battery!

BEAR SAYS

In the event of an emergency, keep cool. Get out of danger, then assess the situation. What are your options? What is the best course of action?

Stay or go?

In a survival situation, one of the most important decisions you will need to make is whether to stay where you are or move on. It is usually safer to stay put. Ask yourself if you will be missed. If so, a search party will almost certainly set out to find you, so it is very important to stay where you are. In a crisis, think about the four basic needs: protection, rescue, water, and food (in that order). If these can be fulfilled, then it's best to stay put.

Moving on

If you are in serious danger, you may need to move. Moving on could also be the best option if no one knows you are missing. If a town or village is in sight, consider going there to ask for help. If you do move on, leave a clear sign, if possible a written note saying who you are, the date and time you left, and where you are headed. Seal the note in a plastic bag to prevent it getting wet or damaged, and leave it in an obvious place, weighted down with a rock.

Lost and found

If you get lost, don't panic. Look around to see if you can spot a landmark such as a road or river to get your bearings. If you have a map, see if you can find the same feature on it, so you can work out your location. If you get separated from your group, it's probably best to stay put and wait for the others to come back and find you.

FIRST AID

There's always a risk of an accident in the wild, so it's important to know how to act in an emergency. This section will show you how to deal with common first aid situations (although it's a good idea to take a first aid class too). Good knowledge, regularly practiced, saves lives.

Bear

IN THIS SECTION:

INTRODUCTION TO FIRST AID

Unfortunately it is very common for people to become unwell or get involved in an accident. If you are keen to go outdoors and explore, it is important to understand some basic first aid. You should make sure an adult is nearby or knows your location at all times, but these handy tips can help keep you safe.

Getting training

If an accident occurs, somebody needs to take charge and decide what to do. You will feel safer and more confident if at least one person in your group has had some first aid training. There are lots of classes available in most towns, and these are often extremely enjoyable as well as teaching a really important life skill.

first aid kit

After training

You may decide that a simple understanding of first aid is enough for you, or you may find that you want to continue learning and join a first aid organization.

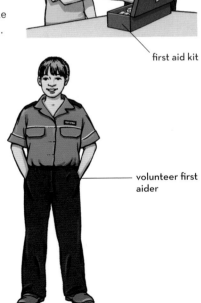

volunteer first aider

First aid kit

It is useful to have at least a basic first aid kit for any outdoor situation. You will need to decide what to bring according to the activities you are planning, the time of year, the length of time you are out, and the needs of the people going with you.

BEAR SAYS

The most effective way to learn first aid is by practicing with others. Ask at your school or library to find out where you can take classes.

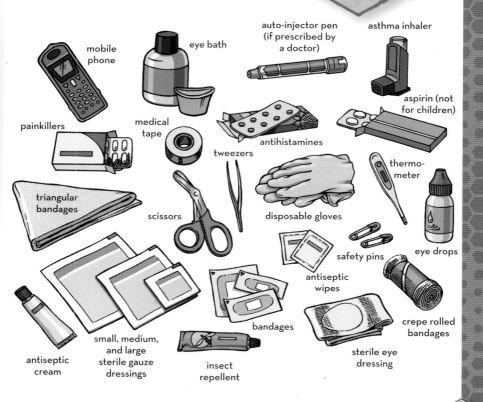

mobile phone

eye bath

auto-injector pen (if prescribed by a doctor)

asthma inhaler

painkillers

medical tape

antihistamines

aspirin (not for children)

tweezers

thermometer

triangular bandages

scissors

disposable gloves

eye drops

safety pins

antiseptic wipes

bandages

crepe rolled bandages

antiseptic cream

small, medium, and large sterile gauze dressings

insect repellent

sterile eye dressing

PREVENTION

If you are planning a trip, it is very important to make a first aid plan so that everyone is clear on what to do if there is a problem. If there are any doubts, it is best to postpone the trip until you feel confident that you can deal with any issues.

Preparation

Preventing first aid situations is far better than dealing with them. Make sure you wear the right footwear and take clothing appropriate for the weather conditions. Bring enough food and water for the time you will be out, and carry a mobile phone in order to call for help if you need it. Make sure you plan your route carefully, and that someone at home knows where you are going and when you expect to return.

Signs of life

It's often quite tricky to tell how serious a situation may be. Doctors spend years training and still sometimes find it difficult, so always call for help—it's better to be overly cautious than delay vital emergency services if someone is hurt.

- Shout loudly if they are unconscious. Use their name and ask if they can hear you.
- Gently shake their shoulders. If they are actually asleep, this should wake them up.
- Tell the injured person you are about to call an ambulance, if they are conscious.
- If someone else is with you, ask them to call the ambulance. Get them to tell you when they have done it or, if possible, do it near you.
- Check if the injured person is breathing. Their chest should go up and down. If you hold a mirror or metal spoon under their nose, their breath will steam it up. Wait at least 10 seconds.
- Check for a pulse on their wrist or the side of their neck—this is tricky and you really need to know what you are looking for. Try practicing on yourself.

How to get help

You cannot help someone else if you put yourself in danger. It is important to check that the situation is safe for you and the injured person. If you can safely take steps to make it safe for them, then do so. If not, wait for help to arrive, but keep reassuring the injured person. Call 911 if the situation is a medical emergency. This is when someone is seriously ill or injured, or their life is at risk.

BEAR SAYS

Stay calm. You are no help to anybody if you panic. It's OK to be scared—brave people are those who do their best even when they are scared.

Calling 911

Once you have made the decision to call the emergency services, it will help if you can tell the operator the following information:

- Your location, including the area or zip code.
- The phone number you are calling from.
- Exactly what has happened. As soon as they know where you are they will start arranging for help to come to you—they may ask for some extra information (this does not delay the ambulance).

Make sure you note down:

- The patient's age, gender, and any medical history.
- Whether the patient is conscious, breathing, and if there is any serious bleeding or chest pain.
- Details of the injury and how it happened. This will help the operator to give you important first aid advice while the emergency staff are on their way.
- If you are in the street, stay with the patient until help arrives.
- Call back if the patient's condition changes or if your location changes.
- If you are calling from a building, ask someone to open the doors and signal where the ambulance staff are needed.
- Lock away any pets or animals if possible.
- If you can, write down the patient's doctor's details and collect any medication that they are taking.
- Tell the paramedics when they arrive if the patient has any allergies.

CUTS AND BRUISES

The most common injuries on any trip are cuts and bruises. These can be quite minor and can be treated using supplies from a standard first aid kit. More serious cuts and bruises may need medical attention.

Hand washing

Before dealing with any first aid situation, make sure you have washed your hands (if there is time and clean water available), and wear gloves to prevent contact with any bodily fluids such as blood, vomit, urine, etc.

Cuts

1. Wash and dry your hands, and wear sterile gloves if possible. Talk to the injured person, tell them what you are doing, and reassure them as you try to stop the bleeding. Press on the area with a clean, dry, and absorbent material for a few minutes. If something is embedded in the cut, leave it there until you get medical advice. You might need to press on either side of it.

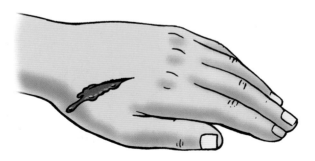

2. If the cut is on a hand or arm, raise it above the patient's head, as this helps to reduce the flow of blood. If a leg is affected, get them to lie down and raise the leg above the level of the heart—you could put their foot on a chair.

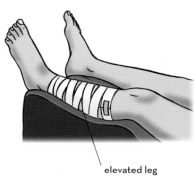

elevated leg

3. When the cut has stopped bleeding, prevent infection by cleaning and drying it, and covering it with a dressing. This may be as simple as washing the cut under the faucet and sticking on a bandage, or it may need to be cleaned with an alcohol-free wipe and covered with a sterile pad with a crepe bandage on top. You can secure the crepe bandage with a safety pin or tape.

black eye

Bruises

A bruise is a bleed that happens under the skin when tiny blood vessels are damaged—usually as a result of a collision. Sometimes a bump also appears—this is just fluid gathering under the skin. As bruises heal, they usually change color. A black eye is a bruise to the eye area.

1. Hold an ice pack on the bruise as soon as it happens for up to 10 minutes. A bag of frozen peas in a towel will work if you don't have an ice pack. If you don't have anything frozen, a clean, damp cloth is better than nothing. If the bruise is extremely swollen, painful, or doesn't go away on its own, medical advice is needed.

2. Someone who has a black eye always needs to be checked by someone with medical expertise, as it is often caused by a bump to the head.

Broken arm or wrist

If you think someone has broken their arm or wrist, look out for the following symptoms:

- Severe pain. They may not want you to touch their arm.
- Their arm may be in a strange position.
- A snapping noise at the time of injury.
- Bruising and swelling.
- Tingling or numbness.
- Difficulty moving their arm.
- In the case of a very serious break, the bone may poke through the skin.

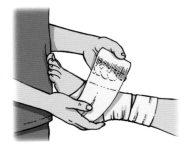

How to identify a break

It is difficult to tell the difference between a sprain and a minor break. Always treat the injury as a fracture until the person has been checked by a medical professional. The patient may also feel dizzy and sick because of the shock.

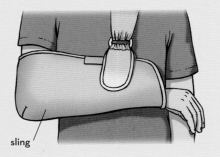

sling

Action

- If it's a bad break, call for an ambulance. Otherwise, go to the nearest emergency department.
- Make a sling that goes around the neck and under the arm. Keep the arm as still as possible.
- Stop any bleeding by pressing on the wound with a clean pad.
- Press an ice pack on the injured area.
- Don't let the patient eat or drink in case they need an operation to fix the broken bone.
- Stay with the injured person. Ideally, one adult will drive and another person will sit next to them in the car.

If someone breaks a limb and their bones are out of place, do not try to straighten them. Call for an ambulance. Broken legs are treated in a similar way to broken arms. Always call for medical assistance.

Sprains and strains

A sprain is the name for an injury to a ligament or tendon. It might have been stretched, twisted, or torn. Ligaments are found in the joints, so the injury might be to a knee, ankle, wrist, or even a thumb.

Strains occur in the muscles and are common in the legs and back. Treatment for sprains and strains is similar.

BEAR SAYS

Usually it is recommended that you move a sprained joint again as soon as it is possible, but a muscle strain may need to be kept still for a few days.

elevated leg

crutches

PRICE stands for...

Protection—use a bandage or support to stop further injury.

Rest—rest the affected joint or muscle. Ask your doctor when you can start moving it again.

Ice—wrap a damp towel around some ice and put it on the area for around 15 minutes every two hours for a couple of days. Don't let the ice touch your skin and don't sleep with the ice in place.

Compression—use an elasticated tubular bandage to stop the swelling, but make sure it isn't too tight, as you don't want to stop the blood flowing. Take the bandage off when you go to bed.

Elevation—keep the limb raised as much as possible.

cold compress

Avoid **HARM** for three days following a sprain or strain:

Heat—don't have a hot bath or go in a sauna.
Alcohol—if alcohol is consumed, the swelling may increase.
Running—don't do any exercise.
Massage—this may slow down the healing.

BITES AND STINGS

Most people who get stung will get better in a few hours or days. However, always treat bites and stings with caution, as some people can have a bad allergic reaction.

Survival tips

- Never disturb bees, wasps, or hornets.
- Don't wear perfume, bright colors, or eat and drink sugary foods if you are in an area with lots of bees, wasps, or hornets.
- Don't swat them or wave your arms around—stand still.
- Wear shoes and avoid loose clothing.
- Keep vehicle windows closed.
- If the stinger is left in your skin, scrape it out with a credit card or similar, using a sideways motion.
- Wash the area, elevate it, and apply a cold compress.
- Painkillers may help. Avoid home remedies.
- Get medical help if you throw up, feel unwell, have been stung on the mouth, throat or near the eyes, or if there is any swelling or breathing difficulties.

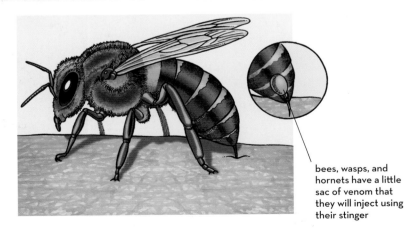

bees, wasps, and hornets have a little sac of venom that they will inject using their stinger

Jellyfish sting survival guide

- Most jellyfish stings are mild and require simple treatment.
- First, get the person out of the water.
- Remove any remaining tentacles using tweezers, and wear gloves if possible.
- Apply a heat pack or put the affected area in hot water.
- Painkillers may help.
- If the person has difficulty breathing, or has been stung in the face or the genitals, call for medical help.
- Vinegar has been shown to stop the box jellyfish from continuing to discharge its stingers. Despite common myths, urinating on the sting is unlikely to help.

✖ BEAR SAYS

Always tell an adult if you've been stung, as some people can have a life-threatening allergic reaction.

use vinegar on
a jellyfish sting

DROWNING

If someone is at risk of drowning, you need to stay calm and act quickly. Get them out of the water before you try to perform first aid. Never put yourself in danger to try and help someone in the water.

Survival guide

Do not put yourself in danger. If the person does not respond to you, do the following:

- Get someone else to call for an ambulance.
- Check their airway and look for signs of breathing.
- If they are breathing but unconscious, put them in the recovery position (see page 68).
- If they are not breathing, give CPR and rescue breaths (see page 72).
- Keep going until they respond to you or help arrives.
- If they start breathing, keep them warm, put them in the recovery position with their head lower than their body, keep checking their breathing and pulse, and make sure they keep talking to you until help arrives.

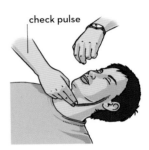

check pulse

tilt head back to clear airway

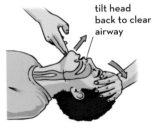

BEAR SAYS

Rescue breaths and CPR are best learned by taking a first aid class and refreshing your knowledge every year, as techniques sometimes change.

FALLS

Falls are extremely common, particularly with toddlers and older people, at home. Even just falling over from standing can cause a fatal injury, so care is essential even if the fall does not appear serious. It is always worth estimating how far a person has fallen, as this can help the emergency services.

Minor trips

If the person has no obvious injuries and can get up, help them to slowly move onto a chair. If you are worried, or they have bumped their head, call for medical assistance. Some symptoms may appear in the hours and days after a fall, so don't leave them alone, and always get them checked by a medical professional if there are any concerns. It is important to find out why the person fell over so it can be prevented in the future, or they can get medical treatment if they have a condition that may cause them to fall again.

paramedics treating the victim of a major fall

Major falls
- Do not put yourself in danger.
- Be extremely careful not to move them if at all possible, as they may have a head or neck injury.
- Follow emergency first aid procedures.
- Ask someone to call the emergency services.
- If they aren't responsive or breathing, give rescue breaths and CPR.
- Try to stop any bleeding.

CPR can be performed anywhere

ELECTROCUTION

If an electric shock occurs in a building, there should be a place to switch off the electricity (usually a fuse box). It is always worth asking an adult to show you where this is located. It can look different from the one in the picture depending upon how old it is.

Survival guide
- Get an adult to switch off the power.
- Don't go near the person or touch them until the power supply is off.
- Follow Dr.' s ABC (see page 70).

the red switches on a fuse box will turn it off

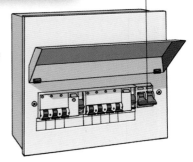

Lightning strike

Being struck by lightning is extremely rare. You can calculate how close a storm is by counting the time between a flash of lightning and a clap of thunder. If it is less than 30 seconds, you need to get to solid shelter. Get inside a building or a car. Avoid touching metal or anything that conducts electricity, and stay away from water. Golf clubs, trees, and umbrellas should also be avoided.

If you can't get indoors, crouch down, making yourself into a small ball with your feet together. Tuck your head in toward your chest and don't lie flat on the ground. Wait until around half an hour after the last flash of lightning before you come out of any shelter, as strikes are common after the storm has passed.

A person who has been struck by lightning may have very minor injuries, or they could be much more severely injured, including the possibility of cardiac arrest.

If the person isn't breathing, begin Dr.'s ABC and deal with any burns or bleeding.

BEAR SAYS

Do not touch anything electrical unless you have been shown how to use it by an adult and they have said that you may use it.

DIARRHEA AND VOMITING

Diarrhea usually clears up on its own in a few days if it's caused by an infection. Make the person drink small sips of water as often as possible, particularly if they are also vomiting. Avoid fruit juice or soda, as they can make things worse. Take care to watch out for signs of dehydration, particularly in young children and elderly people.

Dehydration

Ideally, adults should have enough water, salt, and sugar. A bag of chips and some diluted juice may help. Ginger is a home remedy that some people find helps with nausea. Women in early pregnancy can suffer from sickness and may find that eating ginger cookies helps them feel better.

ginger cookies

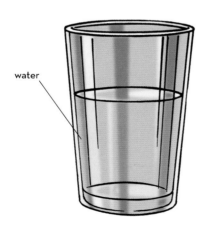

water

RECOVERY POSITION

This is used when someone is unconscious but breathing, and showing no sign of any other life-threatening condition. If you suspect someone has a spine or neck injury, do not move them into the recovery position—wait for the emergency services to arrive.

Why is this helpful?

The recovery position helps to stop the patient's tongue from blocking their airway, and also allows blood or vomit to drain away from their airway safely. It is the safest position to put someone in if you have to leave them.

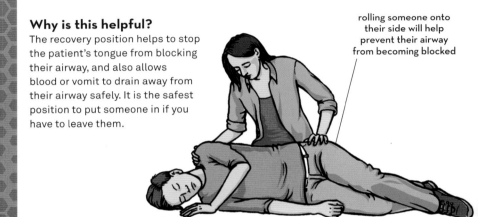

rolling someone onto their side will help prevent their airway from becoming blocked

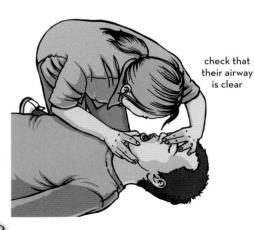

check that their airway is clear

BEAR SAYS

It is useful to practice putting someone into the recovery position. You can find video guides online to learn how to do this.

Moving someone into the recovery position

- If they are lying on the floor on their back, kneel beside them.
- Take their arm nearest to you and place it at a right angle to their body, with their hand pointing up toward their head (step 1).
- Cross their other arm over their chest and place their hand under the side of their head closest to you, with the back of their hand under their cheek (step 2).
- Bend their knee that is farthest away from you.
- Roll them carefully onto their side by pulling their bent knee towards you (step 3).
- Make sure their airway is open.
- Stay with them until help arrives (step 4).

Step 1
Place their arm nearest to you at a right angle, with their hand pointing up.

Step 2
Place their other hand next to their head, with the back of their hand under their opposite cheek.

Step 3
Raise their knee that is farthest from you and use it to pull them toward you onto their side.

Step 4
Make sure their airway is open and stay with them until help arrives.

DR.'S ABC

The Dr.'s ABC is a good way of remembering the steps you need to take when checking a casualty.

Survival guide

Danger.
You must keep yourself safe—you cannot help if you are injured too.

Response.
Ask the person their name or tell them to open their eyes to see if they are conscious. It's OK to shout.

talk to or shout at the person to try for a response

Shout for help.
Get someone else to call the emergency services while you carry out the first aid.

Airway.
Make sure their airway isn't blocked and is open. If they are unconscious, tilt their head back and lift their chin.

check that their airway is clear

Breathing.
Look, listen, and feel for signs of breathing for 10 seconds. If the person is unconscious but breathing normally, they should be placed in the recovery position. If the person is unconscious and not breathing, call for an ambulance and start CPR.

check for signs of breathing

Circulation.
Check for a pulse on their wrist or the side of their neck for 10 seconds. Look for signs of bleeding—don't worry about minor cuts. Press down on any bad cuts with a clean pad and raise the cut above the heart if possible. Keep watching the injured person and look out for signs of shock.

apply pressure to any cuts

CPR

• •

CPR (Cardio Pulmonary Resuscitation) is carried out when somebody falls unconscious and stops breathing completely, or if their breathing is not normal.

Why is it important?

It is important to carry out CPR in order to help prevent brain damage. It is unlikely to restart someone's heart, but it will give the patient a better chance of recovery if the heart can be restarted with a defibrillator.

If you are in a situation where CPR is needed, you will get help over the telephone while the emergency services are on the way. If you are confident, you can carry out CPR with rescue breaths, but otherwise you can just do hands-only CPR, which is also known as chest compressions.

BEAR SAYS

CPR is really hard work and gets tiring fast. It is best if several people can take turns so that the CPR doesn't stop.

Hands only

1. Put the heel of your hand on the breastbone at the center of the patient's chest. Put your other hand on top and lock your fingers together.
2. Make sure your shoulders are above your arms.
3. Use your body weight to press straight down on their chest by about 2 inches.
4. Keep your hands on their chest but relieve the pressure, allowing your hands to come back to their original position.
5. Repeat this about twice per second until the ambulance arrives.

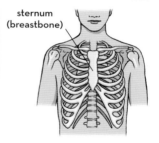

sternum (breastbone)

Rescue breaths

For an adult:

- Give two rescue breaths after 30 chest compressions.

- Tilt their head and lift their chin. Seal your mouth over their mouth, and blow steadily and firmly into their mouth for about a second. Make sure their chest rises. Do this twice.

- Continue this cycle of rescue breaths and chest compressions until they recover or emergency help arrives.

BEAR SAYS

I cannot stress enough the value of taking a first aid class to learn this, or making sure you are supervised by an adult with first aid training.

The instructions are slightly different for children, and different again for infants under one year. You can learn this in specialized first aid classes.

SHOCK

Shock is a life-threatening condition that happens when the body experiences less blood flow than it should. It is a completely different thing from an emotional shock.

Symptoms of shock

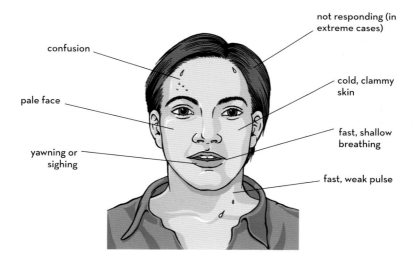

- confusion
- not responding (in extreme cases)
- pale face
- cold, clammy skin
- yawning or sighing
- fast, shallow breathing
- fast, weak pulse

Treatment
- Lay them down, with their legs raised if possible.
- Call for an ambulance.
- Loosen any tight clothing.
- Keep them warm and calm.
- Keep checking their breathing, pulse, and level of response.
- If they become unresponsive, move on to Dr.'s ABC.

ALLERGIC REACTIONS

Always ask people you are traveling with if they are allergic to anything. Make sure they have enough medication for the trip, and that everyone knows where it is and how to deal with any problems.

Food allergies

Nuts, fruit, shellfish, eggs, and cows' milk are commonly associated with food allergies. Allergic reactions can be life-threatening but are often mild. A food allergy could cause an itchy mouth, throat, or ears, swelling, a rash, or vomiting. Anaphylaxis is life-threatening and the symptoms are swelling of the mouth, difficulty breathing, lightheadedness, and loss of consciousness. Some people with food allergies have an auto-injector pen (EpiPen®) that contains a hormone called adrenaline that can be used in emergencies.

Food allergies

- A person with a food allergy should try and prevent a reaction by avoiding any food they know they are allergic to.
- Call for an adult and/or medical help even if the symptoms are mild or have stopped.
- Dial 911 and explain that you think someone is having a severe allergic reaction, and tell the operator what you think has caused it.
- If the person has medication such as an auto-injector adrenaline pen (EpiPen®), help them to use it.
- Place them in a comfortable sitting position, leaning slightly forward to help their breathing.
- If they become unresponsive, open their airway and check their breathing.
- If they aren't breathing, CPR will need to be performed until medical help arrives.

peanuts

auto-injector pen

SPLINTERS

A splinter is when a small fragment of something, most often wood, gets stuck under your skin. It is normally quite a minor injury, but can become infected if not treated properly.

Small splinters

Make sure you have clean hands, then clean the wound with water. If it doesn't hurt, the splinter will work its own way out if it is left alone. If it hurts, you can gently touch the area with tape and see if that pulls it out.

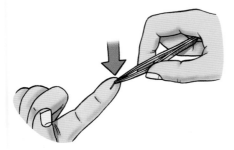

Larger splinters

Clean some tweezers with alcohol. If you can see the end of the splinter, grip it with the tweezers and pull it out in a straight line. Squeeze the wound to make it bleed slightly, as this will help remove dirt. Wash and dry the wound and stick a dressing on if it is needed.

BURNS, HEATSTROKE, DEHYDRATION, AND SUNBURN

Burns can be very serious, and must be treated as soon as possible. The effects of the sun can also be serious, and it is important to take precautions when outside for long periods of time to avoid heatstroke or sunburn.

Burns

1. Keep yourself safe.
2. Stop the burning by removing the person from the area, putting water on the flames, or smothering the flames with a blanket.
3. Remove any clothing or jewelry that is close to the burned area, but don't try to take off anything that is stuck to the skin.
4. Run the burn under lukewarm or cool water. Never use ice or any greasy substances.
5. Keep the person warm.
6. Cover the burn with plastic wrap or a clear plastic bag.
7. Get an adult to provide a suitable painkiller.
8. Sit them upright if the face or eyes are burned.

lukewarm water

BEAR SAYS

Even minor burns should be checked by a doctor or nurse, especially if the patient is under five or over sixty.

cover a burn with plastic wrap

Heatstroke and heat exhaustion

Heatstroke is very serious, but not as common as heat exhaustion. It occurs when a person's temperature becomes very high and their body cannot lower it without help. It can be fatal, so if you think someone may have heatstroke you need to call an ambulance. Heat exhaustion occurs when a person becomes too hot and starts to lose water or salt. They may develop heatstroke if they don't get treatment fast enough. The symptoms of heat exhaustion include tiredness and weakness, dizziness, low blood pressure, headaches, sickness, muscle cramps, heavy sweating, and extreme thirst.

Treatment for heat exhaustion

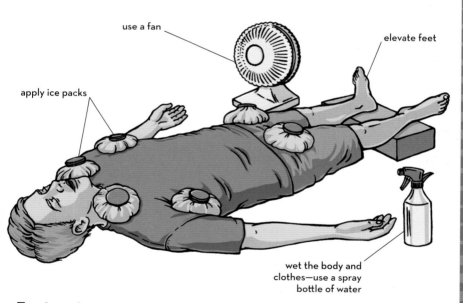

use a fan

elevate feet

apply ice packs

wet the body and
clothes—use a spray
bottle of water

Treatment

Someone with heat exhaustion needs to lie down in the shade. Remove as much clothing as possible, cool their skin with water (you could wrap them in a wet sheet), fan their wet skin, and give them plenty of water, diluted fruit juice, or a sports drink. If they don't start to get better within 30 minutes, call for medical help.

Dehydration

Dehydration happens when someone loses more fluid through sweating, vomiting, diarrhea, or urinating than they can take in from drinking. It is important for everyone to drink plenty of water, but especially if they are exercising more than normal, are in a hot place, have a fever, have diarrhea or are vomiting, or are elderly or very young.

Treatment

Drink plenty of water, take a suitable oral rehydration solution (you can buy this in packets from a drugstore), rest, and massage cramped muscles. If the person doesn't get better fast, get them to a doctor.

oral rehydration solution

Recipe for homemade rehydration solution

You will need: 2 pints of clean water, 6 level teaspoons of sugar, ½ level teaspoon of salt.

1. Add sugar and salt to the water.
2. Be careful not to add too much sugar, which can make diarrhea worse, or too much salt, which is harmful to children. If in doubt, add some extra water.

BEAR SAYS

Sunburn is usually preventable by making sure you stay out of the sun, covering up, and wearing sunscreen. Planning is key.

hat

sunglasses

Sunburn

If a person shows any signs of sunburn, move them to the shade, or preferably indoors. They should take a cool bath or shower, and then apply after-sun lotion. Ask an adult to give them a suitable painkiller if they need it. They need to drink plenty of water, and keep an eye out for signs of heat exhaustion or heatstroke.

UV clothing

sunscreen

CHOKING

Choking occurs when an object, often food, gets stuck in someone's throat, making it difficult to breathe. Choking can be very serious and must be treated as soon as possible.

BEAR SAYS

It is really important not to give abdominal thrusts to pregnant women, or babies under the age of one.

Mild choking

For adults and children over the age of one, encourage them to keep coughing. Then tell them to spit out whatever is causing the problem. Don't put your fingers in their mouth.

Treatment for children under the age of one is different, and it is best to take a first aid class with an expert to learn how to respond to this situation.

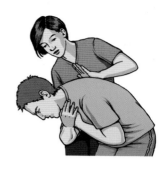

Severe choking

This is when the person cannot speak, cry, cough, or breathe. This information is for dealing with adults who aren't pregnant and children over the age of one. More care needs to be taken with younger children.

- Stand behind the person and slightly to one side. Use one hand to support their chest. Lean the person forward.
- Give up to five sharp blows between the person's shoulder blades with the heel of your hand.
- Check if the blockage has cleared.
- If not, give up to five abdominal thrusts—also called the Heimlich maneuver (see below).
- If the person's airway is still blocked after trying back blows and abdominal thrusts, call 911 and ask for an ambulance.
- Continue with the cycles of five back blows and five abdominal thrusts until help arrives.
- If the person becomes unconscious and isn't breathing, carry out CPR.

Heimlich maneuver

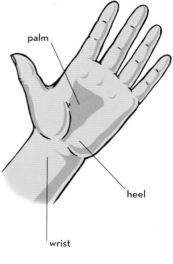

palm

heel

wrist

EYE INJURIES

Any injuries to the eye must be taken very seriously and treated with caution, otherwise you may risk permanent damage to the eye and loss of sight. Even a minor injury should be checked by a doctor or nurse.

Eye wound
- Lay them down and hold their head steady.
- Tell them to keep both eyes still—either look at a fixed point or close them.
- Place a clean pad over the injured eye, then use a bandage to hold it in place.
- Get medical advice.

eye washing

Small foreign body in eye
- Avoid rubbing the eye.
- Sit them down facing a light.

clean pad

- Make sure your hands are clean, then gently open their eyelids with your thumbs and get them to look left, right, up, and down.
- If you can see something, ask them to blink a couple of times to see if that dislodges it (but don't keep blinking for too long, as it may scratch the eye).
- If it still hasn't gone, wash it out with clean water—pour it over the inner corner of the eye (the side nearest their nose).
- If the foreign body doesn't wash out, or it still hurts, get medical advice.

Chemicals in eye
- Flush the eye with large amounts of water, and keep going until medical help arrives, if necessary. This depends upon the chemical, but if in doubt, keep going.
- Call for medical assistance and tell the operator what the chemical was, if possible.
- Keep the patient warm and reassure them—eye flushing can be uncomfortable.

BEAR SAYS

Do not attempt to remove a large foreign body from the eye. Don't press on it—call for medical help.

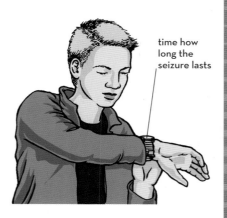

SEIZURES

Seizures can take several forms—a person may go stiff, lose consciousness, fall to the floor, or jerk about. They may just appear to be daydreaming. There are lots of behaviors that could be a type of seizure, and some are more major than others. A person might lose control of their bladder or bowels, or they might bite their tongue or the inside of their mouth.

Treatment

- Protect them from injury. If possible, move furniture and harmful objects away, but don't move the patient unless they are in danger.
- Cushion their head, but don't hold them down. Don't put anything in their mouth.
- Many people with epilepsy wear jewelry or carry a card that will tell you what to do.
- Time how long the seizure lasts.
- Put them in the recovery position once any jerking has stopped.
- Keep them calm, and don't let them eat or drink until they have fully recovered.
- Call for an ambulance if you know it is their first seizure, if it lasts longer than five minutes, if they are injured, or if you feel they need medical help.

time how long the seizure lasts

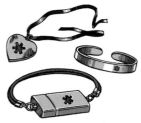

epilepsy jewelry

EMERGENCY MEDICINES FROM NATURE

It's always a good idea to carry a basic first aid kit even on very short trips, but sometimes a medical situation occurs when you have nothing with you. Nature can sometimes provide a temporary remedy, but should be used with extreme caution as there is a risk of making things worse.

Sphagnum moss—bleeding

Sphagnum moss (peat moss) was collected and cleaned on a large scale to be used as a dressing for wounds during World War I. It can absorb 20 times its own volume in blood, and helps prevent infection (as long as it is clean in the first place).

sphagnum moss

Aloe vera—minor burns

After the burn has been cooled and cleaned, the sap from an aloe vera plant can be used to soothe pain and help the skin heal. It cannot be used on anything other than a very minor burn or sunburn, and all burns should still be checked by a medical professional. You can also buy aloe vera in a tube.

aloe vera

Sweet basil—insect repellent

Early Greeks and Romans thought basil would only grow if you shouted and cursed while you planted the seeds. It was used as a remedy for snakebites and scorpion stings. It is thought to repel insects, and some people keep it in their closets to keep moths away from their clothes.

basil

BEAR SAYS

You need to be absolutely sure you know what you are doing with plants. When you step out of the house, take a small first aid kit with you—it makes all the difference.

Making a stretcher

You need: a blanket, two long poles, and a roll of duct tape.

1. Lay the blanket flat on the floor.
2. Place the first pole in the middle of the blanket and fold the blanket over.
3. Place the second pole on top of the blanket, about 25 inches away from the first pole.
4. Fold the blanket over the poles, making sure it overlaps. Bigger blankets work better because there will be more overlap.
5. Wind the duct tape around the poles to keep everything secure (carefully made stretchers can work even without the tape).
6. When lifting a person on the stretcher, raise the head end first, then the feet.
7. When lowering, put the feet end down gently first.

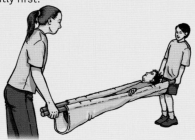

HYPOTHERMIA, FROSTBITE, AND SNOW BLINDNESS

There are many injuries and problems that can be caused by cold weather, and it's important to be aware of the best ways to treat them to prevent serious injury or shock.

Snow blindness

This is a painful eye condition caused by too much exposure to the sun's rays. Symptoms may include watery or bloodshot eyes, twitching, headache, pain, and fuzzy vision. Most commonly, eyes can feel gritty.

snow goggles

Treatment

If you experience snow blindness, go inside and sit in a dark room. Keep your eyes closed, and put something over them to prevent all light from entering the eyes. Then get medical advice. Snow blindness is easy to prevent by wearing suitable sunglasses or goggles. If you have lost your goggles, you could make some by cutting slits in cardboard, to limit the amount of light getting to your eyes.

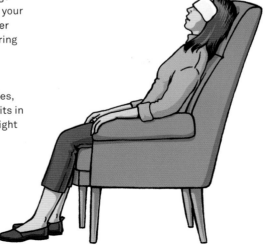

Frostbite

Frostbite occurs when parts of the body freeze due to low temperatures. It is most common in fingers and toes. It can cause permanent loss of feeling in that part of the body, or the tissue can die and become gangrenous.

BEAR SAYS

Look out for symptoms of hypothermia, as a person with frostbite is likely to have hypothermia at the same time.

What to look out for:

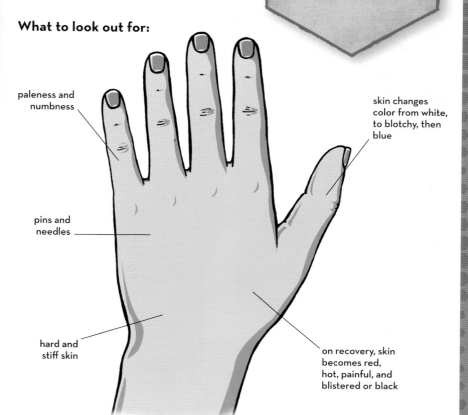

paleness and numbness

skin changes color from white, to blotchy, then blue

pins and needles

hard and stiff skin

on recovery, skin becomes red, hot, painful, and blistered or black

BEAR SAYS

All of these cold weather conditions can be prevented by taking and wearing the correct equipment. Make sure you always follow the advice of an expert and don't take risks.

Frostbite first aid

- Get them to put their hands in their armpits.
- Move them somewhere warm.
- Don't rub the affected area.
- Place the affected area in warm but not hot water (about 104°F).
- Dry the area carefully and apply a light dressing.
- Raise the area above their heart to keep swelling to a minimum.
- Ask an adult to give them some suitable painkillers.
- Get them to a hospital or call for medical help.

Hypothermia

This occurs when the body temperature drops below 95°F, and is a very serious medical condition.

igns of hypothermia

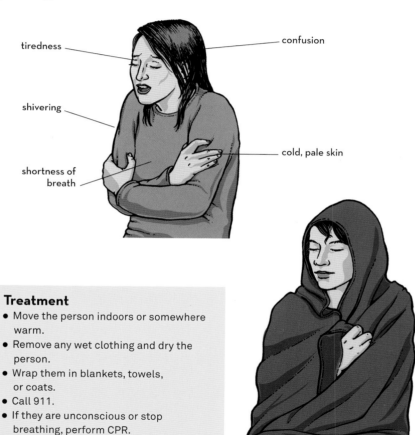

tiredness

confusion

shivering

cold, pale skin

shortness of breath

Treatment

- Move the person indoors or somewhere warm.
- Remove any wet clothing and dry the person.
- Wrap them in blankets, towels, or coats.
- Call 911.
- If they are unconscious or stop breathing, perform CPR.

HEAD AND NECK INJURIES

All head injuries should be treated as serious, as the brain is easily damaged. You should also assume that someone with a head injury also has a neck (spine) injury.

Things to look out for:

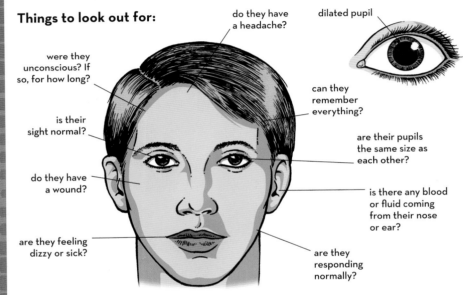

- do they have a headache?
- dilated pupil
- were they unconscious? If so, for how long?
- can they remember everything?
- is their sight normal?
- are their pupils the same size as each other?
- do they have a wound?
- is there any blood or fluid coming from their nose or ear?
- are they feeling dizzy or sick?
- are they responding normally?

First aid
- Sit them down.
- Hold something cold on the injury.
- Treat any wounds by pressing to stop the bleeding.
- Perform Dr.'s ABC, CPR, and rescue breaths if needed.
- Call for medical help—all head injuries need to be checked, even if they don't appear serious.

Neck injury

A person with a neck injury might have numbness, weakness, pain, or tingling in their arms or legs. They may have a sore neck, back, or head. Their neck or back may be at a strange angle, and they may also have a head injury.

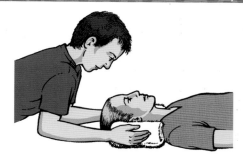

First aid

- Keep them calm and make sure they don't move.
- Kneel behind their head, rest your elbows on something to keep your arms steady, and grip either side of their head. Keep their head, neck, and spine in a straight line, but don't cover their ears.
- Keep this support until the emergency services arrive.
- You could use rolled up towels on either side of their head if there is someone who can help you.

Log roll

A log roll is a technique that is sometimes carried out by several people on someone with a spinal injury to prevent them from choking if they throw up. It should only be done if absolutely necessary and you need to be trained before attempting it.

BEAR SAYS

Be very careful not to move a person with a neck or spine injury, as this can cause paralysis.

POISONING AND INTOXICATION

Poisons can occur in nature, but can also come from different chemicals you may find in your house. If you think someone has swallowed a poisonous substance, get medical help immediately.

First aid

- Get them to sit still and stay with them.
- Get them to spit out anything still in their mouth.
- Take off contaminated clothing and wash their skin with water, taking care not to get the poison on you.
- If they are unconscious, try and wake them up to spit.
- If you can't wake them up, put them in the recovery position.
- Wipe any vomit from their mouth, and keep their head pointing down.
- Don't let them eat or drink anything, don't put your hand in their mouth, and don't make them vomit.
- Perform Dr.'s ABC, CPR, and rescue breaths if needed.
- Tell the emergency services as much as you can. For example, what they swallowed, when they swallowed it, how much they swallowed, and whether it was an accident or on purpose.

Alcohol poisoning

If someone has drunk too much alcohol, stay with them, as they are more likely to have an accident or injury because their thinking is impaired. If they lose consciousness or are vomiting, get medical help, as they may be suffering from alcohol poisoning.

First aid

- Keep them sitting up and awake, and stay with them.
- Give them some water and help them drink in small sips.
- Put them in the recovery position if they are unconscious.
- Keep them warm.
- Always get an adult to help, as intoxicated people can be unpredictable and sometimes violent.
- Don't give them coffee or put them in a cold shower. Keep them still.
- If you are worried, call for medical help.

HEART ATTACKS AND STROKES

A heart attack is a medical emergency that happens when the blood supply to the heart muscle is suddenly blocked. Call for an ambulance immediately, as a heart attack can be fatal. A person having a heart attack may have lots of different symptoms, including a sharp pain in their chest, or pain traveling out from the chest into the jaw, arms, stomach, neck, or back. Symptoms are different for everyone, so if in doubt always call for medical help.

First aid
- Sit them down and make them comfortable.
- Ask an adult to decide if they should slowly chew a 325 mg aspirin tablet (children shouldn't take aspirin, and some people are allergic to it, so be careful).
- Some people have a condition called angina, and they may have a spray or pills for this. Get it for them and give it to them if they ask.
- Check their breathing until help arrives.
- Perform Dr.'s ABC, CPR, and rescue breaths if needed.

aspirin

Stroke

A stroke happens when the blood supply to part of the brain is cut off. A person having a stroke needs an ambulance as fast as possible, as they will make a better recovery if they get hospital treatment quickly.

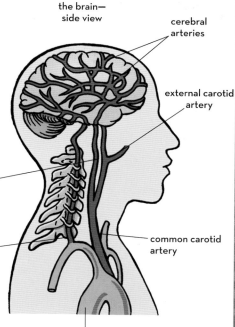

the brain—side view

cerebral arteries

external carotid artery

common carotid artery

blood supply from the heart

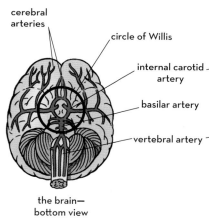

cerebral arteries

circle of Willis

internal carotid artery

basilar artery

vertebral artery

the brain—bottom view

Signs of a stroke
1. **F**ace: Look at their face—is it lopsided? Can they smile? Has one eye or one side of their mouth dropped?
2. **A**rms: Can they lift both of their arms and keep them up? Is one weaker than the other?
3. **S**peech: Can they talk normally? Is their speech garbled or slurred? Can they talk at all?
4. **T**ime: It is time to call an ambulance if you notice any of these symptoms. Tell the operator that you suspect the person is having a stroke.

BEAR SAYS

We can reduce our risk of having a heart attack or stroke by eating healthily and getting plenty of exercise.

SIGNALING

If you are lost, injured, or the weather takes a turn for the worse, knowing how to use the resources around you to find help could be a lifesaving skill. Once you have mastered these important signaling skills, you're ready to get out there and start adventuring!

Bear

IN THIS SECTION:

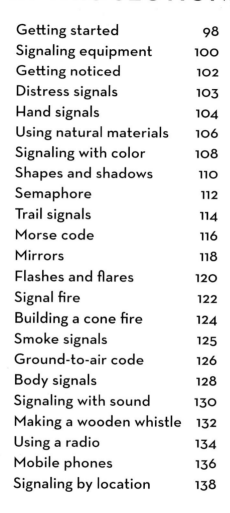

GETTING STARTED

Signals are communication methods. They may work at close range or over quite long distances. Signaling is fun to learn and practice, but it's also a vital survival skill. In an emergency, it could save your life.

Emergency signals

This section explores signals that are used by explorers, survival experts, and all who love the outdoors. Knowledge of signals allows you to call for help in an emergency. This could be vital if you are injured, lost, trapped by bad weather, or survive a disaster such as a plane crash.

Main types of signals

There are two main types of signals: visual signals that can be seen, and audio signals that can be heard. These correspond to the main human senses: sight and hearing. Visual signals include flashes, flags, hand and body signals, handwritten notes, and mobile phone texts. Audio signals include whistle blasts, phone calls, and radio transmissions.

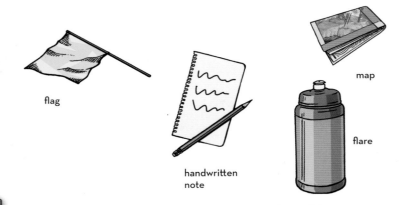

flag

handwritten note

map

flare

What to take

Practice signaling outdoors in an open space. The right clothing and equipment are important on all outdoor trips and expeditions. Wear or take several layers of clothing so you can put on a layer if you are cold or take it off if you are hot. Take an anorak in case of rain.

sunscreen

watch

gloves

food/drink

first aid pouch

sun hat

warm hat

walking boots
or shoes

BEAR SAYS

Before all expeditions, tell an adult where you are going and what time you expect to get back. If you are very late, the adult can then raise the alarm.

rucksack

compass

SIGNALING EQUIPMENT

You don't need a lot of fancy gear to learn the basics of signaling. Start with a few simple items and add more gear gradually as you need. Some items are high-tech, while others can be made or improvised cheaply.

Basic signaling kit

Many of these items are vital to wilderness survival in general. Carry them with you whenever you venture into the wild.

mobile phone

pouch

matches

flashlight

whistle

mirror

pencil

notebook

waterproof bags

markers

Advanced signaling gear

Special items of signaling equipment
are also available.

flint and steel can be
used to light fires

flagging tape
can be used to
mark trails

heliograph—a
special signaling
mirror

flare (see
pages 120–121)

fluorescent marker panel
can be used to send
ground-to-air signals

radio transmitter (see
pages 134–135)

personal locator
beacon (see pages
134–135)

binoculars are useful
for reading signals at
a distance

satellite phones can be
used to communicate from
most locations

GETTING NOTICED

Signaling is about getting noticed—sending messages that stand out in your environment. Distress signals are used in emergencies to alert rescuers that you need help.

Attracting attention

The key to getting noticed is: Bigger, Brighter, Different.

Bigger—Humans are very small compared to the great outdoors. Make your signals large if you want to be noticed!

Brighter—Brightly colored clothing, flashes, flares, fires, and loud sounds stand out in nature.

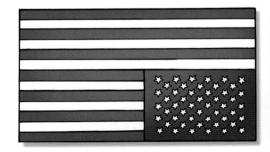

Different—Clothing and man-made objects with straight lines look out of place in nature. Anything that looks wrong, such as an upside-down flag or raised car trunk, will also attract attention.

DISTRESS SIGNALS

Distress signals are used to call for help in an emergency. Rescuers will risk their own lives to answer distress calls. These signals are taken very seriously, so should never be misused.

SOS

SOS is an internationally recognized distress call. Traditionally used by ships' captains, it is now used anywhere. SOS can be written as letters or sent as Morse code (see pages 116–117).

Mayday

This distress call was originally used by airmen in trouble. Traditionally sent by radio, it comes from the French "*m'aidez,*" meaning "help me." It is said three times.

Three for danger

Any signal repeated three times is an internationally recognized distress call. This includes three blasts of a whistle, three light flashes, or three fires, arranged in a line or triangle.

BEAR SAYS

If your distress call is not answered immediately, don't panic. It may take some time to attract attention. Wait about a minute between sending distress signals in a set of three.

HAND SIGNALS

Hand signals are used by soldiers on maneuvers. They are fun to use if you want to communicate silently with friends outdoors. Use them to move through woods or open spaces without attracting attention.

Practice the signals with your friends first, to make sure you all understand the gestures and how to give them clearly. You could also develop your own signs.

BEAR SAYS

Hand signals are useful when stalking wildlife to avoid frightening animals. Here you should give signals slowly—any sudden movements will spook wildlife.

Movement hand signals

come here

hurry

stop

obstacle

meet here

go here

wait

Bear Grylls

Actions hand signals

listen

look

I don't understand

cover this area

I understand

breach

Numbers

one

two

three

four

five

six

seven

eight

nine

ten

USING NATURAL MATERIALS

Natural materials such as rocks, pebbles, and branches can be used to spell words and make visual signals. You can also draw letters in snow, mud, and sand.

Choose your location

Picking the right location for your signal is important. It depends whether you want the signal to be seen from the air or from the ground.

Good locations include:

1. A hilltop or ridge with an all-around view can be seen from the air and from all sides.
2. A signal on steeply sloping ground can be seen from below.
3. A clearing in a wood or forest is visible from the air.
4. An open grassy space can be seen from the air or higher ground.

Gather materials

Decide on the materials to use, depending on what can be found in the environment. You could use largish rocks, branches, logs, or seaweed or pebbles on a beach. Carry materials to the chosen location.

Prepare the ground

Clear the ground of debris such as rocks that could distract attention or confuse the viewer.

Scaling up

Make letters as large as possible, and at least three times as tall as they are wide. Letters should be at least 10 feet tall, 3 feet wide, and 3 feet apart to be seen from the air.

BEAR SAYS

If you move away from your signal location, leave an arrow to mark which way you have gone.

choose a patch of level or gently sloping ground

Writing letters

You can trace letters with a stick in damp sand, mud, or snow. You can also tramp down snow with your feet, or shovel it away to expose dark soil beneath. On a beach, write a message above the high tide mark so it won't wash away.

SIGNALING WITH COLOR

Bright colors are great for attracting notice. In an emergency, a bright flag or garment can alert rescue services or even someone just passing by.

Make a flag
You can improvise a flag from brightly colored clothing, a space blanket, bivy bag, or life jacket tied to a stick. Wave the flag above your head if help is in sight.

  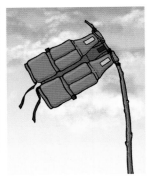

Make a scarecrow
Make a cross of sticks and pull a T-shirt over it like a scarecrow. Remember: Bigger, Brighter, Different— this signal will really stand out.

Tinsel tree

Shred a space blanket and tie the strips to a tree to signal in an emergency.

BEAR SAYS

Once the emergency is over, be sure to erase all distress signals. Remember that rescuers will put their own lives in danger to save yours.

Flagging tape

Strips of flagging tape can be tied to trees or bushes to attract attention or mark your route. You can write on this tape with markers.

Dye marker

Some survival kits contain packets of dye marker. This colored powder creates a clear signal if spread on water, and is also effective on snow and sand.

SHAPES AND SHADOWS

Shapes, shadows, and silhouettes can be used to send visual signals. Remember that straight lines and sharp angles stand out well in natural surroundings.

Creating shadows

Shadows can be used to spell out letters when it's sunny. Create shadows by piling snow, sand, or earth into walls to form letters. The walls should be at least 1 foot high, and will stand out even better if you dig a trench to make the wall.

You can also pile up branches or rocks to form letters, but the letters will need to be very well defined for the shadows to be legible.

Message cairns

Cairns can also be used to leave written messages. Place the note in a plastic bag, and leave it under the top stone of the cairn.

Inuit cairns

In the Arctic, the Inuit traditionally build human-shaped cairns called inuksuit (singular: inuksuk) to mark trails and herd caribou.

Silhouettes

An unusual outline silhouetted on a ridge will attract attention. Build a cairn of stones to attract notice or mark your route.

BEAR SAYS

Your knowledge of distress calls could save someone's else's life if you notice an emergency signal no one else has seen.

Marker panels

Some survival kits contain a marker panel. This is a large square of cloth or canvas with different colors on the front and back. The regular, man-made shape stands out in nature. Fold the cloth to send different signals as shown.

Improvised marker panel

You may be able to improvise a marker panel using a tent and groundsheet or a bivy bag and space blanket.

walking in this direction

need equipment

need medical attention

need first aid supplies

land here

do not land here

color code

☐ white
☐ yellow
☐ blue

SEMAPHORE

Semaphore is a system of signaling using flags to spell letters and words. Semaphore was widely used by sailors in past centuries and is still used today.

Improvised flags

Improvise flags with brightly colored cloths or clothing. If you tie or sew the cloth to sticks, your signals will be clearer. Red and yellow stand out well at sea. Blue and white flags are clearly visible on land.

Sending semaphore messages

Practice semaphore with a friend in an open space. Hold each position for at least six seconds. Count seconds by saying "one thousand and one, one thousand and two…" slowly. Then move to the next position, making the change obvious.

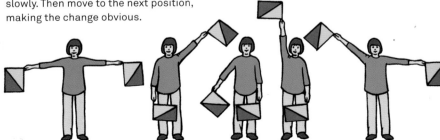

Receiving messages

Binoculars can help you read signals over a distance. When receiving messages, draw the arm positions. Work out the letters and words after the signaler has stopped sending.

A

B

C

D

E

F

G

H

I

J

K

L

M

N

O

P

Q

R

S

T

U

V

W

X

Y

Z

cancel

numbers follow

BEAR SAYS

Hold a flag in each hand with your arms out straight. There are eight possible positions for each arm.

TRAIL SIGNALS

Trail signs are used by scouts to mark routes and communicate with others following. In an emergency, trail signs show rescuers which way you headed if you moved on.

Practice trail marking

Hone your survival skills with friends by taking turns to lay a trail and follow it. You can also develop your own signs, known only to your group.

BEAR SAYS

Marking the trail is very useful if you are hiking in unfamiliar territory. On the way back, or if you lose your way, simply follow the markers to retrace your steps.

Laying a trail

Each sign can be made using sticks, stones, or tufts of grass. You can also scratch signs in mud or sand. Leave trail markings at regular intervals, and at all junctions where followers will be unsure which way to go.

this way

gone home

turn left/right

go straight on

message hidden (10) paces this way

not this way

water this way

message this way

message this way (over obstacle)

group split up

MORSE CODE

Morse code is an international code made up of short signals, or "dots," and longer signals, or "dashes." Developed in the 1830s, it is named after American inventor Samuel Morse, who invented the telegraph.

How Morse code works

Different combinations of dots and dashes represent numbers and letters of the alphabet. String letters together to make words and sentences.

Morse code messages can be sent using visual signals such as smoke, flags, and light flashes, or you can use audio signals such as whistle blasts or beeps on a radio.

whistle

flashlight

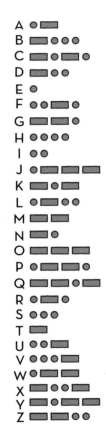

Sending Morse code

Practice sending and receiving Morse code messages with a friend. When sending, keep messages short and simple. Write down the message, then look up the code.
Send slowly, making signals as clear as possible. Pause slightly between each letter, with longer pauses between words.

Receiving

When receiving, write down the code for each letter. Work out the message when the signal stops.

BEAR SAYS

SOS in Morse code is three dots, three dashes, and three more dots. Memorize this simple signal—in an emergency, it could save your life.

Morse code in flags

You can signal in Morse code using a flag or cloth tied to a stick. Hold the flag upright. Move it to the right for dots and to the left for dashes. Make dashes slightly longer than dots. To signal over a distance, move the flag in figures of eight to the right or left. Over a short distance these exaggerated movements aren't necessary.

MIRRORS

Mirrors reflect sunlight. You can use them to send light flashes that can be seen from a great distance. You can also use a mirror to signal in Morse code.

Improvised mirrors

A hand mirror can be used to send a signal. If you don't have a mirror, other shiny objects will also reflect light.

aluminum foil

CD

tin cup or plate

glass shard

Practice sending and receiving Morse code messages with a friend using a hand mirror. It's not easy but, with practice, your technique will improve.

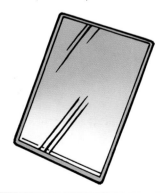

BEAR SAYS

Keep moving the mirror slightly while aiming the flash at an aircraft, so you attract the pilot's attention without blinding him or her. Never signal an aircraft except in dire emergency.

Signaling aircraft

In an emergency you can signal an aircraft using a mirror. This method takes little effort, but only works when it's sunny.

1. Hold the mirror at shoulder height and point toward the sun.
2. Stretch your other arm out with two fingers up and palm facing inward. Sight the aircraft (or another target) between your fingers.
3. Angle the mirror so a spot of reflected light hits your fingers.
4. Lower your outstretched hand while keeping the mirror at the same angle to direct the flash at the plane.

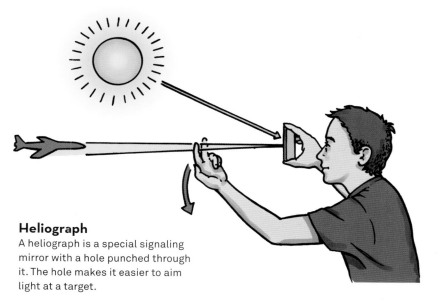

Heliograph

A heliograph is a special signaling mirror with a hole punched through it. The hole makes it easier to aim light at a target.

Using a heliograph

Hold the mirror up to your face and sight the aircraft through the hole. Your reflection shows a spot of light falling on your face. Tilt the mirror so the light spot disappears through the hole while you are looking at the aircraft.

FLASHES AND FLARES

The signals described so far in this book mainly work in daylight. Flashlights can be used to signal in darkness, while flares can be seen during the day or at night.

Flashlights

Flashlights can be switched on and off to attract notice or send Morse code signals. Strobe lights flash automatically to draw attention.

A flashlight taped to a branch and waved above your head sends a clear and obvious signal if rescue is in sight.

Three for danger

Three flashlights flashed at once sends a clear distress signal.

Whirling glow stick

A glow stick can be attached to a string and whirled to create a conspicuous circle of light.

Conserve batteries

In an emergency, save flashlight batteries by signaling at intervals.

Flares

Handheld flares produce a bright light or plume of colored smoke that can be seen for miles. There are two main types: pistol flares and rocket flares. Warning: flares are very dangerous and need to be handled with great care.

Handling flares

Keep flares dry and well away from a campfire. Remove one flare at a time from the box and replace the lid.

Lighting a flare

Flares can get very hot so wear gloves if possible. Read all instructions carefully. Hold the flare out at arm's length. Point it at a 75° angle, not directly upward, so burning debris does not fall on you or your camp. Follow the instructions to ignite the flare.

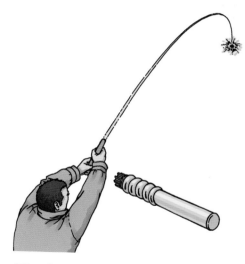

BEAR SAYS

Position yourself in an open space that can be seen from all sides. Point the flare well away from anyone else. Brace yourself for the kickback when the flare ignites.

Warning

Many flares are still burning when they hit the ground. Flaming debris can burn a hole in an inflatable life raft in a split second. Avoid using flares in very dry areas where you could start a fire.

SIGNAL FIRE

Fire and plumes of smoke are visible by day or night. For centuries, fire has been used to send signals, often in emergencies.

Fire location

Choose a sheltered spot on level ground. If possible, make a platform of green (freshly cut, sap-filled) sticks or branches to keep the fire off the ground. You could arrange large logs or rocks around the fire to form a windbreak.

Build a pyramid fire

A pyramid fire can be ignited quickly if rescue is in sight. Place a ball of tinder in the center and pile kindling over it. Place small fuel sources over the kindling to form a pyramid shape.

Methods of fire lighting

The easiest way to light a fire is using matches, a lighter, or a flint and steel kit. Rub the steel along the flint to create a spark. If you have none of these, you can ignite tinder by focusing the sun's rays using a magnifying glass or even a pair of glasses.

Blow gently on an ember to produce a flame. Add more fuel gradually, to avoid smothering the fire.

flint and steel

matches

magnifying glass

lighter

BEAR SAYS

Direct the spark or flame at the tinder. If it's windy, use your body as a windbreak, and cup your hands around the flame.

Warning

Fire is very dangerous, so you need to be careful. Only light a fire with adult supervision. Have a bucket of water or sand handy to put out the fire if needed.

BUILDING A CONE FIRE

These techniques create signal fires that are clearly visible in open country or dense woodland.

Cone fire

You need: green sticks or branches of various sizes, knife, string.

1. Sharpen one end of three long, straight sticks. Bind the blunt ends loosely with string, rope, or wire.
2. Fan out the sharpened ends to form a tripod shape. Push the sharpened ends into the ground.
3. Tie three shorter sticks about 6 inches from the base of the tripod to form a triangle.
4. Lay smaller, straight sticks onto the triangle to form a platform.
5. Build a pyramid fire on the platform and place green branches over it. Leave a gap to ignite the tinder in the center when you hear or see rescue coming.

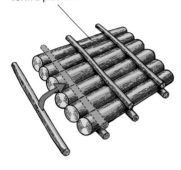

use small, straight sticks to form a platform

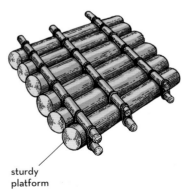

sturdy platform

completed cone fire

SMOKE SIGNALS

A plume of smoke acts as a distress beacon. You can also send an SOS signal by fanning fire with a cloth.

Three fires in a line or triangle produce a clear distress signal.

Fire raft

Dense forest or jungle hides fire and smoke. A river, pool, or lake provides an open space from where a fire will be visible, so the best place to site a signal fire is on a raft.

Build a raft

Gather straight branches or bamboo poles. Lash them together crosswise using string or rope. Tie the raft to both banks of a river and build a pyramid fire on top.

BEAR SAYS

Make sure any fire is completely extinguished before you move on. The smallest spark could start a destructive blaze, which could put you and others in danger.

GROUND-TO-AIR CODE

Ground-to-air code is a method of signaling aircraft in an emergency. These signals can be made using natural materials.

Gather materials

Rocks, branches, beach pebbles, or bright objects can all be used to make the symbols. You can also write the code in mud, sand, or snow.

Ground-to-air code is particularly useful as it makes it clear to aircraft if you need to be rescued, or if you need supplies to be dropped.

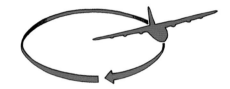

Clearing a space

Choose open ground, a summit, or ridgetop to site the signal. Clear away debris such as sticks and rocks that could confuse the message. If using natural materials, make sure they stand out well against the background.

Make the symbols as large and clear as possible. These symbols should be five or six times as tall as they are wide.

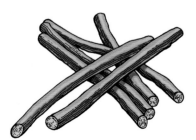

BEAR SAYS

The word FILL helps you to remember the three most important signals: food and water, injury, and all well.

require
doctor—serious
injuries

require compass
and map

all well

require medical
supplies

go this way

no

unable to
proceed

I am going
this way

yes

require food
and water

probably safe
to land here

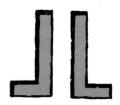

I don't
understand

BODY SIGNALS

Body signals are also used to signal from the ground to aircraft. In an emergency, these signals can be used to guide a helicopter to a landing.

Ground signals

In an emergency your first contact with the outside world is likely to be a search aircraft. Make this contact count by learning standard ground-to-air signals. You can use objects, as well as your own body, to seek help.

Key

1. Need medical assistance
2. Use drop message
3. No
4. Yes
5. Do not attempt to land here
6. Land here
7. All OK, do not wait
8. Pick us up
9. Need mechanical help
10. Our receiver is operating
11. Wait, I can proceed shortly

Prepare a helicopter landing zone

You need an open space at least 130 feet across. The ground should be fairly flat and even. Remove debris such as rocks and branches, and also cardboard, paper, and plastic that could blow around.

BEAR SAYS

Exaggerate the body positions. Note that some positions are made sideways to the aircraft. Use a cloth to make the signals for "yes" and "no" clearer.

helicopter will approach downwind if possible

mark the edges of the touchdown area with bright items

clear an area of at least 60 ft. in diameter—the ground should be as flat as possible

a garment tied to a stick shows the pilot which way the wind is blowing

stand outside the touchdown zone where you are visible

Warning

Beware of the helicopter blades and powerful downdraft. Approach the helicopter from the front, never from the rear.

SIGNALING WITH SOUND

The human voice doesn't carry far, and if you shout for a long time you will get hoarse. Noisemakers such as whistles and drums produce louder sounds that travel farther.

Objects that make noise
These objects are commonly used to create a sound signal.

banging two objects together

shouting

car horn

starter pistol

Improvise
Improvise a drum by beating a hollow log or metal sheet with a stick. Or strike a metal cup or plate with a stick or spoon. Claves can be made with two dry, hardwood sticks. Strike one stick against the other to produce loud clicking sounds.

BEAR SAYS

Whistles help members of a group keep in touch with one another, for example, when moving through mist or woodland. You can work out your own signal code.

Whistle

Blowing a whistle uses far less energy than shouting, and the sound carries farther. Three or six blasts of a whistle is an emergency signal. Or send SOS in Morse code: three short, three long, and three short blasts.

Grass squeaker

Put a blade of grass between your thumbs and knuckles as shown. Hold it taut, purse your lips, and blow between your thumbs.

Wolf whistle

1. Stretch your lips over your teeth. Cover your teeth completely, but keep your lips relaxed.

2. Put two fingers between your lips as far as the first knuckle, with your fingertips pointing toward your throat.

3. Flatten your tongue against the bottom of your mouth until it's almost touching your gums.

4. Blow forcefully so air passes over your tongue and bottom lip through your fingers. Adjust the position of your lips and fingers until you produce a whistle.

MAKING A WOODEN WHISTLE

Make your own whistle using elder wood or bamboo. Fun and easy to make and use, this could also save your life in an emergency.

You need:
a knife, stick of bamboo or elder, or any wood with a central pith, thinner sticks.

1. Cut a straight length of elder or bamboo about the width of your finger.

2. Cut a finger-length section and remove the bark. Push out the central pith using a smaller stick.

3. Cut a 45°-angle notch about an inch from one end as shown.

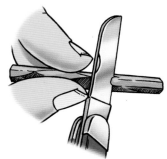

4. Cut a length of straight stick slightly thicker than the central hole, and pare it down to make a dowel to fit inside.

5. Shave one side of the dowel slightly to make it flat. Insert it into the mouth end as far as the notch, so the flat side lines up with the notch. Trim off any end.

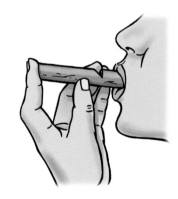

6. Test the whistle by blowing while blocking the far end with your finger. Adjust the dowel to alter the pitch.

7. Block up the far end using the dowel but without the flat side. This must be airtight.

USING A RADIO

Radio provides an easy way to send and receive complex messages. Learn how to operate a radio before heading off into the wild.

Radio equipment
Lightweight two-way walkie-talkies are an excellent way of communicating with a partner, a group, or base camp. If on an expedition, agree to specific times a day for transmitting messages.

Getting a signal
Radio range is limited to within line of sight of a receiver. Reception is often poor in forests and cities. Climb to higher ground to get or improve a signal.

Prepare to send a message
1. Raise the aerial but hold it level with the ground to get the best signal. Keep the aerial away from your body, clothing, or the ground.
2. Check if the microphone is activated by voice or button. Have the message ready and tune into the right frequency.

BEAR SAYS
If reception is poor you can send an SOS in tones and beeps. The button may be labeled Key, CW, or Tone. SOS is three beeps, three tones, three beeps.

Transmitting a message

Hold the microphone 4 inches from your mouth and speak slowly and clearly. In an emergency, say "Mayday, Mayday, Mayday." Give your name and position. State the nature of the emergency and the number of people. Then say "Over" and wait for the reply.

International distress frequencies

VHF Radio: Channel 16
CB radio: Channel 9
Family Radio Service UHF: Channel 1
Amateur (ham) radio: 2182kHz, 14.300MHz, 14.313MHz
Airband Radio: 121.5MHz, 243MHz
UHF Radio (Australia): Channel 5

PLBs

Personal Locator Beacons (PLBs) are small devices resembling handheld radios. In an emergency, activate the device to send a distress signal with your position to a satellite. This is then relayed via a ground station to a search and rescue center which will send help.

MOBILE PHONES

Mobile and satellite phones present the easiest form of communication—provided that you have a signal. It's also vital to conserve batteries or keep them charged.

Mobile phones and smartphones

Mobile phones allow you to call and text while on the move. If you are lost or injured you can take and send photos of locations or injuries. Smartphones provide access to the Internet. You can download maps or a GPS app, which shows your exact location.

Getting a signal

As with radios, mobile phones only work in line of sight of a transmitter. They may not work in hills, canyons, or off the beaten track. Heading uphill may improve the signal. If the signal is weak, text rather than call.

No signal?

If you have no signal but a full battery, send a text and keep the mobile phone on while you move: the text will be sent if you enter an area with reception. If there's no signal and your battery is low, turn off and text from likely spots such as hilltops.

Satellite phones

Satellite phones allow you to call and text from any location via satellites. Survival experts use them because they work in wilderness areas.

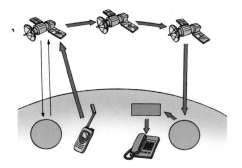

Conserving batteries

Radios, mobile phones, and satellite phones all rely on batteries, which need to be kept warm and dry. In icy weather, keep them warm in a pouch next to your skin. Turn off devices when not in use, or buy a solar charger so you can charge your phone without electricity.

International emergency numbers

Call these numbers only in true emergencies. Get ready to give your name, location, and the nature of the emergency.

911	US, Canada, Mexico
112	Europe except UK, New Zealand, India
999	UK
000	Australia

Phonetic (spoken) alphabet

Use the phonetic alphabet to make sure any letters you use during phone calls or radio transmissions are understood.

Letter	Code Word	Letter	Code Word
A	Alpha	N	November
B	Bravo	O	Oscar
C	Charlie	P	Papa
D	Delta	Q	Quebec
E	Echo	R	Romeo
F	Foxtrot	S	Sierra
G	Golf	T	Tango
H	Hotel	U	Uniform
I	India	V	Victor
J	Juliet	W	Whiskey
K	Kilo	X	X-ray
L	Lima	Y	Yankee
M	Mike	Z	Zulu

BEAR SAYS

Before going abroad, find out local emergency numbers and load them onto your phone.

SIGNALING BY LOCATION

Different types of terrain suit various forms of communication.
Here's a handy summary of what will work where.

Open country and/or grassland

Visual signals such as flares, flags, flashes, and fire work well, particularly from hills and ridges. Phones and radios function well except in remote areas.

Forest, woodland

Wooded areas are poor environments for signaling. Vegetation hides flashes, flags, and even fire and smoke, and blocks radio and phone signals. Don't ignite a fire or flare in a dry forest. Sounds such as shouts and whistle blasts are also masked by vegetation. Seek clearings, high ground, or open water to send signals.

Desert

Open desert terrain suits visual signals—provided there is anyone to see them. Flags, flares, flashes, fire and smoke, and even written messages are easily spotted by aircraft. If lost or stranded, stay in the shade or by a broken-down vehicle, which will be conspicuous from the air.

At sea

Marine areas have their own signaling methods. Lighthouses and buoys mark dangerous waters. Ships communicate using radio, satellite phones, lights, and flags. Use a light, whistle, flare, or dye to call for help from a small boat or in the water. Take great care when igniting flares. Helicopters will use a winch to rescue survivors.

Mountains

Canyons and valleys block visual, radio, and telecom signals. Climb to ridges or summits above the tree line to signal using flashes, flares, flags, fire, or body signals. In mountain rescue code, six light flashes or whistle blasts signal an emergency. The response is three flashes or blasts.

Polar regions

Visual signals work well here. Bright colors, man-made shapes, flares, fire, smoke, and dye stand out well against snow. Batteries and even whistles can freeze if not kept warm. Air rescue is likely to come from a plane with skids, not a helicopter.

TRACKING

There's nothing quite like watching animals in their natural habitat, but tracking animals isn't easy. For this, you'll need to hone your detective skills. Learn how to stay hidden in the wild, identify animals from their footprints, and heighten your senses to become the best animal detective. Enjoy the adventure!

Bear

IN THIS SECTION:

GETTING STARTED

Tracking is the art of detecting animals from the clues they leave behind. This ancient skill allows you to experience the natural world as never before. In a real-life survival situation, you need tracking skills to stay alive!

Ancient craft

In prehistoric times, our ancestors were skilled trackers who survived by hunting animals. In some parts of the world, hunter-gatherers still track animals today.

Survival skills

Survivors of disasters such as plane crashes use tracking skills to find food to stay alive.

Sensing danger

Tracking skills can alert you to dangers in the wild. Rodent or insect tracks show you need to store food out of the reach of animals. Tracks of large animals such as bears show you need to be very wary!

Tune in to nature

It takes patience to become an expert tracker, but the rewards are great. As you become skilled, you will be able to detect animals in the wild and even in parks and cities.

Clothing

Prepare for all sorts of weather on tracking expeditions. You will often be moving slowly, so bring an extra layer of clothing, and cover up if the sun is hot. You will need a warm hat or sun hat, waterproof jacket, gloves, sweater, several layers of clothing, and boots or walking shoes. Take these with you on expeditions: a drink and snack, sunscreen, first aid kit, flashlight, and mobile phone. A foam sit mat can also be useful.

BEAR SAYS

Always tell someone where you are headed when you go tracking. You can also send a text or leave a note.

Equipment

These items are useful for tracking: notebook, pen/pencil, binoculars, magnifying glass, camera, ruler or tape measure, sealable plastic bags for collecting samples, tweezers, flashlight.

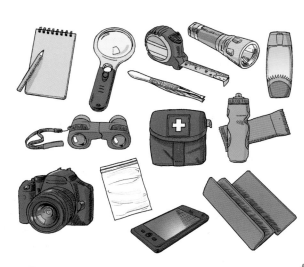

BASICS OF TRACKING

Tracking is about becoming aware of your environment. Animals have keen senses. To become an expert tracker, you need to fully use your senses too.

Sight is the most important sense for humans. Train yourself to notice small details and traces of animals.

Scent can alert you to the presence of animals, and tell you whether clues, such as droppings, are fresh.

Hearing lets you listen for bird and animal calls, and sounds such as splashing, crashing, and rustling leaves.

Touch tells you which way the wind is blowing. Feeling the texture of hair and wool can help you identify animals.

BEAR SAYS

Develop night vision by allowing your eyes time to adjust to darkness. Use your flashlight as little as possible, or tape clear red film over the glass.

Hearing   Sight

Touch   Scent

Wide-angle vision

Many animals have eyes set far back on their head. This gives a wide field of vision, so they can spot danger from all sides.

Broaden your vision

Develop wide-angle vision in an open space. Hold your arms out in front of you. Scan the area between your hands. Move your arms wider apart and scan again. Keep moving your arms apart to increase your field of view.

Train your hearing

Small sounds such as squeaks and grunts can reveal hidden animals. Ask a friend to hide a wristwatch in a room at night. Train your hearing by following the ticking to find the watch.

Hone your observation

Train your powers of observation by playing this memory game. Ask a friend to put a collection of small objects on a tray, and cover with a cloth. Remove the cloth for a minute and memorize the objects. Replace the cloth. How many objects can you remember?

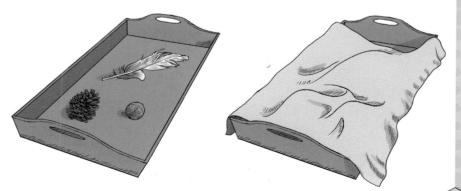

UNDERSTANDING ANIMALS

To track animals effectively you need to understand their needs and put yourself in their position. Expert trackers can work out what an animal was doing just from its tracks.

Fight for survival

Wild animals face a daily struggle for survival. They must find food and water while avoiding enemies. For a species to survive, animals must breed.

Migration

Animals such as caribou travel on long, regular journeys called migrations. They migrate to find food, avoid cold, or reach a safe place to rear their young. Many birds migrate.

Water

Almost all animals need to drink clean water daily. Desert animals travel a long way to reach water.

Food and feeding

Animals have varied feeding habits. Deer, antelope, and rabbits are herbivores, or plant eaters. Carnivores such as leopards hunt other animals for food. Shrews and hedgehogs are insectivores, or insect eaters. Bears and pigs are omnivores—they eat a variety of foods.

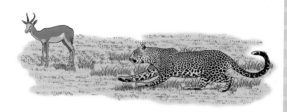

Habitats

All animals are suited to their habitat. Penguins could not survive in a woodland, and woodland animals would die in Antarctica. Knowing which animals live in local habitats will help you to identify tracks.

Daily rhythms

Animals such as gray squirrels are diurnal—mainly active by day. Nocturnal creatures such as raccoons find food at dawn, dusk, or under cover of night.

Senses

Animal senses are tuned to their way of life and habitat. Cats have good night vision and keen scent and hearing. Touch-sensitive whiskers help them to hunt at night.

Range

Animals such as wolves are widespread. Wolves live on the tundra and in forests, mountains, and other wild places throughout the Northern Hemisphere.

TRACKS AND FOOTPRINTS

Expert trackers are skilled detectives. The smallest clues provide vital evidence. Trails and footprints are two of the main types of evidence that animals are around.

Where to look

Animal tracks show up best in muddy banks of streams, rivers, lakes, and ponds as well as on damp sand on the seashore. Soft, dusty ground and freshly fallen snow also show trail marks. If claw marks are clearly visible, the prints are likely to be fresh.

otter paw print

gull print

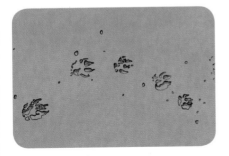

lion paw print in dust

wolverine print in snow

Animal trails

Animals wear down paths along regularly used routes, for example, from the den to a favored feeding spot. Look for these trails through woods or long grass, particularly near gaps in fences. Note that trails may be used by more than one animal.

BEAR SAYS

After rain, dew, or frost, animal tracks show up as dark lines through grass.

grass bent in one direction shows which way the animal was headed

where deer cross a road, they leave a trail of soil and leaves on the tarmac

deer nibble bark and leaves at head height—this creates a strip bare of vegetation, called a browse line

FEET AND MOVEMENT

Animal feet vary greatly according to species. Animals also move in different ways, leaving distinctive tracks which help to identify species.

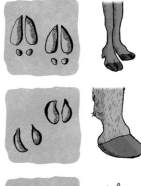

Mammal feet
Mammal feet come in different shapes.

Dogs, cats, and foxes have neat, rounded paws which leave rounded prints.

In deer and cattle, two toes have evolved into a cloven (split) hoof. The print is called a slot.

Horses walk on just one toe, which has evolved into a hoof.

Touching the ground
Bears, badgers, otters, shrews, and hedgehogs walk on the soles of their feet. The print is of the whole foot.

Dogs, cats, and foxes walk and run on their toes. The print shows toe pads but not the heel.

Gait

Animals also move at different speeds. Each gait leaves a distinctive track.

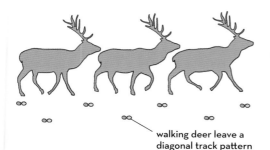

walking deer leave a diagonal track pattern

BEAR SAYS

You are unlikely to find perfect sets of prints. Build up a picture of the animal by studying partial prints.

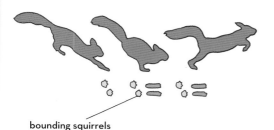

bounding squirrels leave a distinctive track

Bounding

Squirrels, hares, and weasels bound rather than run. The hind and front feet contact the ground alternately.

Bird prints

Birds' feet may be slender, webbed, or lobed. Most birds have three toes pointing forward and one backward. Birds move in different ways: songbirds hop, pigeons and pheasants walk, while ducks and geese waddle.

KEEPING RECORDS

Keep a record of all the tracks and clues you find in a notebook or journal. Your notes, sketches, and photos build up into an amazing picture of wildlife in your area.

Tracking journal
This sample page shows how to keep records in a notebook.

Describe the find and take measurements. Add a sketch or photo. A careful drawing is often clearer than a photo, because you can show details clearly and leave out any debris that's not part of the print.

Note the date, time, location, and conditions.

3 inches

Date: Jan 25
Time: 2 pm
Location: Bank of stream in Coopers Wood
Conditions: Damp and muddy
Five prints. Four neat toe pads
with claws plus rear pad
Species: Fox
Activities: Drinking or
patrolling territories

Can you identify the animal? What do you think it was doing?

Taking measurements

Use a ruler or tape measure to record the length and width of prints. Use a tape measure to record stride length and the width of the animal trail.

Tracking stick

You can use a tracking stick to record animal tracks. All you need is a straight stick and a few rubber bands. Place one end of the stick level with the top of the print. Slide a rubber band down the stick to mark the heel. Use more rubber bands to record the stride length and the trail width. You can also use a marker instead of rubber bands.

BEAR SAYS

Use resealable plastic bags to collect finds such as bones, hair, and feathers. Add paper labels to identify your finds.

OTHER RECORDS

Plaster casts and tracings provide an excellent record of the prints you find. Make a tracking bed to record perfect prints.

Make a plaster cast

You will need:

plaster of Paris

bottled water

spoon

mixing bowl

trowel

old toothbrush

paints

1. Bend a piece of cardboard into a ring large enough to enclose the print. Secure with a paper clip.

2. Press the ring into the soil around the print.

3. Spoon plaster of Paris into the mixing bowl. Gradually add water and stir until the mixture is thick and smooth.

4. Pour the mix into the ring. Fill to near the top. Gently tap the edge of the ring to bring any bubbles to the surface.

5. Wait about 15 minutes for the mix to harden, depending on conditions. Then gently lift or dig the cast from the soil.

6. When dry, brush to remove dirt, or clean with an old toothbrush and water.

7. Paint the cast to highlight the print. You can also varnish it.

Making tracings

You need a clear sheet of acetate and a
permanent marker. Place the acetate
over the print and carefully trace the outline
with the pen. Alternatively, you can use a non-
permanent marker to trace onto a small piece
of acrylic. Scan or copy the drawing then wipe
clean and use again.

Making a tracking bed

Make a tracking bed near an animal burrow
by clearing loose twigs, rocks, and leaves
away. Smooth the ground flat with the edge
of a ruler. Return the next day to find a clear
set of prints.

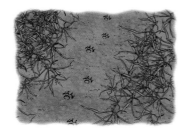

Making a backyard tracking bed

Make a tracking bed in your backyard by sprinkling sand onto
a tray to a depth of about an inch. Smooth flat with the edge
of a ruler. You could put a spoonful of dog or cat food in the
center to act as bait. Leave overnight and check the next day.

ANIMAL HOMES: NESTS

Birds build nests in spring to rear their young. Each species builds a characteristic shape. Mammals such as squirrels and dormice also make nests.

What's it made of?

Birds use different materials to build their nests. Shape, structure, and materials all help to identify nest builders.

Warning sign: never approach birds' nests in spring and summer, when they may contain eggs or young.

BEAR SAYS

Size is an obvious clue as to the identity of nest builders. The bigger the bird or animal, the larger the nest.

woodpeckers

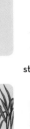

storks

wood pigeons

coots

kingfishers

ringed plovers

1. Songbirds weave twigs into a cup-shaped nest. Thrushes line their nests with mud.

2. Wrens build ball-shaped nests with a small, round entrance hole. Long-tailed tits build similar nests made of moss and lichen bound with cobwebs.

3. Wood pigeons build loose, untidy nests of sticks.

4. Rooks nest high in a clump of trees. The nest colony is called a rookery.

5. Storks build large, untidy stick nests perched on chimneys.

6. Peregrine falcons nest on crags and church steeples.

7. Kingfishers nest in holes along the river bank.

8. Swallows build nests of mud and saliva under the eaves of houses.

9. Coots and grebes build floating nests of vegetation.

10. Ringed plovers nest in hollow scrapes on the beach. Their eggs resemble pebbles.

11. Woodpeckers chisel out nest holes in tree trunks using their sharp beaks.

12. A squirrel nest, called a drey, is built into the crook of a tree.

peregrine falcons

squirrels

songbirds (thrushes)

rooks

swallows

wrens

DENS AND BURROWS

Mammals such as foxes and badgers shelter underground in dens and burrows. It's not always easy to identify who is living underground. Many insects, some reptiles, and even birds live in burrows too.

badger sett

BEAR SAYS

Droppings and food scraps near the burrow can help to identify the owner. The size of the entrance hole provides another clue.

bear den

hare form

prairie dog burrow

1. A fox's home is called a den. Smelly droppings and bones mark the entrance.
2. A badger's den is called a sett. This underground home has several chambers. A large heap of earth marks the entrance. Badgers sharpen their claws on a nearby tree.
3. A rabbit warren contains a network of tunnels. Look for small, round droppings and nibbled grass near entrance holes.
4. Hares shelter above ground in grassy hollows. This hidden home is called a form.
5. Bears hibernate in rocky caves or hollow trees in winter. Never approach a bear's den.
6. Mole hills of excavated earth show where a mole lives underground.
7. A prairie dog colony has many chambers linked by tunnels. Burrowing owls, ferrets, and rattlesnakes may move into these burrows.
8. A squiggly heap called a worm cast marks the entrance to an earthworm's burrow. The heap is made of undigested soil that has passed through the worm's body.
9. The nests of meadow ants create grassy mounds in open pasture.
10. Wood ants nest in large mounds of pine needles in woodlands.

rabbit warren

mole hills

meadow ant mound

worm cast

fox den (earth)

wood ant nest

POOP AND PELLETS

Animal poop, or scat, is another sure sign that animals are about. Look closely and you may see evidence of what the animal has been eating. Skilled trackers can identify animals from their scat.

Whose poop?

Animals produce droppings of different shapes. Droppings contain dangerous bacteria, so never touch them with your hands—use a stick if you want to investigate. Always wash hands thoroughly after any contact with poop, or wipe with a moist tissue.

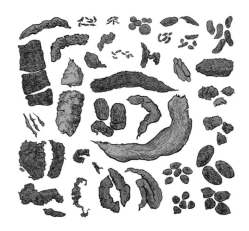

Animals such as foxes, otters, and badgers use dung and urine to mark their territories. Badgers dig small holes called latrines. Otters leave their scat, called spraint, in an obvious place such as on a rock.

Pellets

Birds of prey cannot digest the fur and bones of their prey. They cough up these remains as pellets. Bits of fur and bones provide clues about the predator's diet.

BEAR SAYS

Are the droppings fresh? Fresh dung is moist and attracts flies. As time goes on, droppings decay or dry out.

Size and shape

Different bird species produce pellets of different shapes, sizes, and colors.

Dissect an owl pellet

Pellets can be dissected to find out what the bird has been eating.

1. Soak the pellet in a tray filled with water for at least an hour. Add a few drops of disinfectant to the water.

2. Tease the pellet apart using tweezers. Separate the bones and clean them in a disinfectant solution.

3. Sort and arrange the bones into skulls, limb bones, ribs, etc. Can you identify the owl's prey?

SIGNS OF FEEDING

Animals leave debris as they feed. These leftovers can give away their whereabouts. Nibbled vegetation shows the presence of plant eaters, while predators leave the remains of kills.

Nibbled nuts

Birds, rodents, and squirrels all eat nuts. Mice nibble small, neat holes in shells. Woodpeckers and nuthatches wedge nuts into tree bark to open them with their bills. Squirrels bury nuts to eat later—a behavior called caching.

Grass and leaves

Rabbits and rodents snip plant stems using sharp front teeth called incisors, leaving neatly sliced edges. Deer lack incisors so they leave frayed edges as they strip twigs and bark.

Thrush's anvil

Thrushes smash snails against rocks to reach the soft mollusks inside.

Pinecones

Squirrels and some birds eat pine seeds. Crossbills pluck the seeds, leaving cones with ragged edges. Squirrels strip pinecones to the stalk.

Beaver damage

Beavers gnaw through saplings to build their underwater home, called a lodge.

Bird kill

Tufts of feathers mark the spot where a hawk has killed a songbird.

Rooting boars

Boars and pigs leave a line of debris as they dig beneath the grass for roots and fungi.

Sap sucker

Woodpeckers leave a line of holes in trees as they dig for sap below the bark.

TINY NIBBLERS

Nibbled leaves, holes in wood, and other debris mark where insects and other minibeasts have been feeding.

Leaf damage

Caterpillars are hungry feeders! Some species nibble large holes in leaves, others munch inward from the edges. Some fly, and beetle larvae "mine" leaves from the inside, leaving a pale trail.

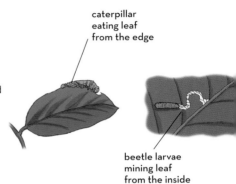

caterpillar eating leaf from the edge

beetle larvae mining leaf from the inside

Ant trails

Look for ant trails leading from the nest to a food source such as fruit or an injured minibeast. Watch the insects scurry to and fro along the trail. Can you see what they are carrying back to the nest? Tropical leafcutter ants snip leaves to carry back to their nest.

Galls

Some wasps and flies lay their eggs in buds, which provide food for the larvae. The plant defends itself by forming a growth called a gall around the insect.

Wood borers

Some beetles lay their eggs in dead wood or lumber. The grubs eat their way through the timber and then fly away as adults, leaving exit holes in the wood.

Butterfly pupa

Young butterflies pass through a pupa (chrysalis) stage as they change from larvae into adults. You may find a pupa case hanging from a leaf or twig by a silken thread.

Cuckoo spit

This unpleasant-looking blob of foam hides a young insect, called a froghopper nymph, as it sucks sap from plant stems.

Spiderwebs

Spiders spin silken webs to trap their victims. Look closely to see prey such as woodlice tied up with silk.

Ant lion pit

Young insects called ant lions dig cone-shaped pits and hide at the bottom. Ants slide down the steep sides into the insect's jaws.

BEAR SAYS

Use a magnifying glass to get a close-up view of clues such as nibbled leaves. Shine a light behind a mined leaf to highlight the insect's trail.

OTHER FINDS

From time to time, keen-eyed trackers will come across finds such as bones, shells, and feathers. These provide clues about animal anatomy and their way of life.

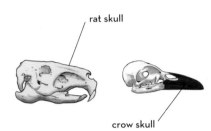

rat skull

crow skull

Skulls
Skulls and other bones can tell you a lot about body structure. Field guides and online sites help to identify species. The shape and size of teeth and jaws reveal an animal's diet.

fox skull

Feathers
Birds molt their plumage at least once a year. Again, online sources and field guides can help to identify the species.

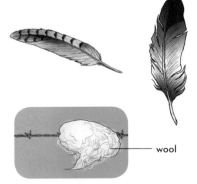

Hair and wool
Hair and wool caught on barbed wire shows where animals pass regularly.

wool

hair

Wing print
A wing print in the snow shows where a bird escaped a predator, such as a fox.

Molted skins
Reptiles, insects, and woodlice molt their skins as they grow. Snakes shed their skin in one piece. This is called sloughing. A dragonfly climbs out of its case to become an adult insect.

By the sea
Shells on the shore are the remains of marine mollusks. This holed shell has been pierced by a predatory mollusk called a whelk. This leathery "mermaid's purse" is the egg case of a dogfish.

Eggshells
Baby birds that hatch naturally usually leave a shell in two neat halves. A ragged hole pecked from the outside is the work of a predator.

Dung ball
Dung beetles roll cattle dung into neat balls and lay their eggs inside.

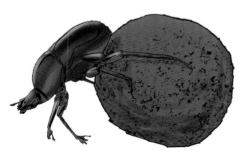

BEAR SAYS

The shape of the skull provides clues about the animal species. Wearing gloves, clean bones with an old toothbrush and disinfectant solution.

STALKING

Stalking is the art of watching animals without being noticed by them. Remember, animals are always on the alert for danger, so avoiding detection takes a lot of skill.

Camouflage
Wear clothing that blends in with your surroundings. If you don't have army-style camouflage, wear natural colors such as greens and browns. Avoid bright colors. Patterns or patches help to break up your outline.

make sure to camouflage your skin if necessary

Skin camouflage
Pale human skin stands out in nature. Smear stripes of mud onto exposed pale skin such as face and hands, or use charred wood from a campfire.

Using binoculars
Keep shiny objects such as camera and binoculars hidden under your jacket, with the strap around your neck. When you need to use binoculars, raise them slowly to your face, keeping your eyes on the target. Avoid sudden movements.

Outline and shadow

Animals can easily detect the human outline and shadow. Disguise your outline with leafy twigs or ivy. Avoid casting a shadow on small creatures such as insects.

Silhouette

Your outline stands out clearly silhouetted against the skyline. Keep low or hide behind cover. Water creatures will spot you immediately if you peer over the bank of a pond or stream.

Using cover

Shrubs and trees provide good cover for watching wildlife. If you have to move through open ground, crawl or crouch low, and use long grass, ditches, or hollows as cover.

BEAR SAYS

The best way to avoid detection is to keep very still and quiet. Find a spot behind cover and be patient. A foam sit mat will keep you comfortable while waiting.

ADVANCED STALKING

Predators use stealth to approach their prey, creeping forward while keeping to cover. You can use similar techniques to get close to wildlife without being seen.

Fox walk

Advance slowly and carefully. With your weight on your back foot, place the ball of your front foot down, testing for twigs or dry leaves that might crack or rustle. If the ground is clear, move your weight forward onto your front foot. Repeat. Advance just a few paces then stop, listen, and look around you. Avoid sudden movements.

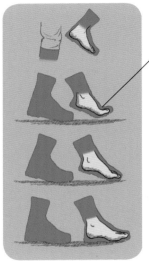

carefully test the ground with the balls of your feet

Freezing

If you are spotted, freeze. If you keep quite still, animals may not be able to spot you, or will see you as less of a threat.

Leopard crawl

Use this crawl to move through long grass. Get down on all fours with your weight on your hands and knees. Move your right elbow and left knee at the same time, then the opposite pair.

Stomach crawl

Use this crawl in open ground. Lie on your stomach with your arms out in front and legs splayed behind. Move forward by pulling with your forearms while pushing with the inside of your feet.

Moving downwind

Your scent will give you away if you approach animals from upwind, with the wind blowing your scent toward them. Test wind direction by licking one finger and holding it up in the breeze. Or drop a few grass stems to see which way they blow. Circle around at a distance until you are downwind of your target.

BEAR SAYS

If you spot an animal, don't try to get too close—watch at a distance. If it moves off, don't follow quickly. After you have finished watching, move away quietly.

MAKING A HIDE

A hide will conceal you while you watch wildlife. Set up your hide near a den or burrow, where many tracks cross, or by a stream where animals drink.

You need: 5–6 canes, camouflaged tarpaulin, string, tent pegs

Camouflage a plain tarpaulin by sewing patches of brown or green cloth onto a plain one. Ask adult permission first.

Pyramid hide

1. Tie the canes loosely at one end with string. Fan out the other ends to make a cone shape. Press the canes into the ground.

2. Drape the tarpaulin around the pyramid shape. Fasten the opening with safety pins or by looping string through the eyelets. Leave a gap at eye height for wildlife watching.

3. Peg the tarp down or secure the edges with rocks or branches. Add a stool or mat inside for comfort.

4. Camouflage the outside with leaves, ivy, or branches.

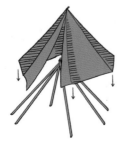

Camouflaged A-frame

1. You can also rig a camouflaged tarp as an A-frame shelter. You need a length of cord. Tie one end of the cord to a tree at a height of about 3 feet. Peg the other end to the ground so the cord is tight.

2. Drape the tarp over the cord. Fold any spare material underneath to act as a groundsheet. Peg or secure the edges with rocks or branches. Use branches to disguise the entrance, leaving a gap to observe wildlife.

BEAR SAYS

Research animal behavior and choose the right time for your stakeout. Get into position at least half an hour before. Sit tight and be patient!

ANIMAL TRACKS: CARNIVORES

The mammal group called carnivores includes dogs, cats, foxes, weasels, and badgers. All of these animals are meat-eating predators that hunt other animals for food.

Domestic dog F, H 2–4 inches
Range: worldwide except Antarctic
Habitat: with humans
Diet: varied, including meat, scraps

Wolf F, H 4–5 inches
Range: North America, Europe, Asia
Habitat: forest, grassland, desert, tundra
Diet: mammals including deer, hares, rodents

Coyote F 2.5 inches; H 2 inches
Range: North America
Habitat: forest, grassland, desert, urban
Diet: rabbits, rodents, reptiles, insects, fruit

Red fox F, H 2 inches
Range: North America, Europe, Asia, introduced to Australia
Habitat: forest, grassland, mountain, desert, urban
Diet: small mammals, birds, insects, scraps

Domestic cat F, H 1–2 inches
Range: worldwide except Arctic and Antarctic
Habitat: with humans
Diet: meat

Lynx F 4 inches; H 3 inches
Range: Europe, Asia, North America
Habitat: forest
Diet: hares, rodents, birds

Brown bear F, H 9–12 inches
Range: North America, Europe, Asia
Habitat: forest, mountain, tundra
Diet: fish, carrion, berries, vegetation, insects, fungi, rodents, sheep

Striped skunk F, H 2 inches
Range: North America
Habitat: forest, grassland, desert, urban
Diet: insects, worms, reptiles, amphibians, birds, eggs, berries, leaves

Raccoon F 2 inches; H 4 inches
Range: North and Central America
Habitat: forest, grassland, mountain, near water, urban
Diet: insects, worms, fruit, nuts, fish, amphibians, eggs

European badger F, H 2.5 inches
Range: Europe, western Asia
Habitat: woodland, meadows, urban
Diet: worms, insects, amphibians, fruit, grain, small mammals, birds

Stoat F 0.75 inches; H 1.5 inches
Range: North America, Europe, Asia, introduced to New Zealand
Habitat: forest, grassland, moorland
Diet: rodents, rabbits, shrews, birds, fish

F—Front
H—Hind

BROWSERS AND GRAZERS

Deer, cattle, sheep, and horses all belong to a group of mammals called ungulates. Deer, cattle, and sheep have cloven hooves. In horses, just one toe has evolved into a hoof. All are plant eaters, as are marsupials such as kangaroos.

Moose F, H 4.5–6 inches
Range: North America, Europe, Asia
Habitat: marsh, woodland
Diet: leaves, twigs, moss, water plants

Red deer F, H 3 inches
Range: Europe, Northwest Africa, Asia, introduced to Australia, New Zealand, South America
Habitat: grassland, woodland, moor
Diet: grass, plants, heather

Roe deer F, H 1 inch
Range: Europe, western Asia, introduced to Australia
Habitat: woodland, grassland
Diet: grass, leaves, shoots, berries

White-tailed deer F 3 inches; H 2.5 inches
Range: North, Central and South America
Habitat: woodland, scrub
Diet: grass, shrub, fungi, lichen, nuts

Caribou/Reindeer F, H 4 inches
Range: North America, Scandinavia, North Asia
Habitat: tundra, forest, mountain
Diet: plants, leaves, twigs, lichen

Muntjac F, H 1–1.5 inches
Range: Asia, introduced to UK
Habitat: forest, farmland
Diet: grass, leaves, shoots

Cattle F, H 4 inches
Range: worldwide except Antarctica
and Arctic
Habitat: farmland, meadows
Diet: grasses

Sheep F, H 2.5 inches
Range: worldwide except Antarctica
and Arctic
Habitat: farmland, moors, scrub
Diet: grasses, plants

Horse F, H 4 inches
Range: worldwide except Antarctica
and Arctic
Habitat: farmland, grassland, mountain
Diet: grasses, grain

Wild boar F, H 3.5 inches
Range: Europe, North Africa, Asia
Habitat: forest, scrub
Diet: leaves, roots, fungi, small
mammals, reptiles, carrion, eggs,
manure

Red kangaroo H 16 inches
Range: Australia
Habitat: grassland and scrub
Diet: grass

Numbat F, H 5 inches
Range: Australia
Habitat: eucalyptus woodland
Diet: termites, ants

SMALL MAMMALS

Rodents such as rats, mice, and squirrels nibble plant food using chisel-like front teeth called incisors. Rabbits and hares also have incisors. Shrews and hedgehogs hunt insects. They are rarely seen, but you sometimes find their tracks.

Brown rat F 0.4 inches, H 1 inch
Range: every continent except Antarctica
Habitat: near water or humans
Diet: seeds, grains, many other foods

House mouse F, H 0.25 inches
Range: worldwide
Habitat: near humans
Diet: seeds, grains, fruit, meat

European rabbit F 1 inch; H 2.5 inches
Range: Europe, North Africa, introduced to Australia
Habitat: grassland, farmland, scrub
Diet: grass, twigs, bark, crop plants

Brown hare F 1 inch; H 2.5–6 inches
Range: Europe, Asia
Habitat: dry grassland
Diet: grass, plants, twigs, bark

Jackrabbit F 1.5 inches; H 4 inches
Range: western North America
Habitat: grassland and farmland
Diet: grass

Gray Squirrel F 1.5–2 inches; H 2 inches
Range: North America, introduced to Europe
Habitat: woodland, urban
Diet: nuts, flowers, buds

Prairie dog F 1 inch; H 1.5 inches
Range: North and Central America
Habitat: grassland
Diet: wood, bark, pine needles, buds, roots,
seeds, leaves

Alpine marmot F, H 4.5 inches
Range: Europe
Habitat: mountain
Diet: grasses, grain, insects, spiders, worms

North American porcupine F 2.75 inches;
H 3 inches
Range: North America
Habitat: forest, brush
Diet: wood, bark, pine needles, buds, roots,
seeds, leaves

Common shrew F 0.2 inches; H 0.5 inches
Range: Europe
Habitat: woodland, grassland, hedges
Diet: insects, slugs, spiders, worms,
amphibians, small rodents

Hedgehog F, H 2 inches
Range: Europe, northwest Asia
Habitat: forest, grassland, gardens
Diet: insects, slugs, fish, frogs, worms, baby
mice, birds

BIRDS

Birds come in many sizes, from hummingbirds to ostriches. They have very different tracks because their feet are suited to their habitat. They also move in different ways, whether hopping, walking, or running.

House sparrow 1 inch
Range: worldwide
Habitat: urban and farmland
Diet: seeds, plants, insects, worms

Song thrush 1 inch
Range: Europe, Asia, North Africa, introduced to Australia, New Zealand
Habitat: woodland, gardens
Diet: snails, slugs, insects, worms, fruit, berries

Raven 2 inches
Range: North America, Europe, Asia, North Africa
Habitat: open country
Diet: rodents, invertebrates, carrion

Wood pigeon 2 inches
Range: Europe, Asia, North Africa
Habitat: woodland
Diet: seeds, plants, invertebrates

Common pheasant 2.5 inches
Range: Europe, Asia, introduced to North America, Australia, New Zealand
Habitat: woodland
Diet: seeds, plants, berries, invertebrates

Turkey 4 inches
Range: North America and introduced worldwide as farm stock except Antarctica
Habitat: woodland, farms
Diet: seeds, berries, nuts, invertebrates

Roadrunner 3 inches
Range: North America
Habitat: open country, desert
Diet: insects, lizards, snakes

Sparrowhawk 2.5 inches
Range: Europe, Asia, North Africa
Habitat: woodland
Diet: woodland birds

Mallard duck 2 inches
Range: North America, Europe, Asia,
North Africa, introduced to Australia,
New Zealand
Habitat: near freshwater and coasts
Diet: plants, snails, insects,
crustaceans, worms

Canada goose 3 inches
Range: North America, Europe, Asia
Habitat: near water
Diet: plants

Coot 1.5 inches
Range: Europe, Asia, North Africa,
introduced to Australia, New Zealand
Habitat: near freshwater
Diet: algae, plants, seeds, fruit, small
animals, eggs

Gray heron 4 inches
Range: Europe, Asia, North Africa
Habitat: near water
Diet: fish and other water creatures

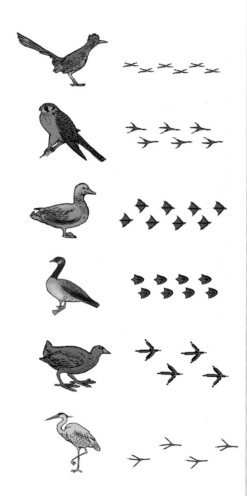

BY WATER

Animal tracks show up clearly on sandy seashores and along the muddy banks of streams, lakes, and rivers. A wide variety of birds, mammals, reptiles, and other small creatures live here.

On the seashore

Herring gull 2.5 inches
Range: coasts worldwide
Habitat: coasts and inland
Diet: carrion, eggs, fish, crustaceans, many other foods

Eurasian oystercatcher 2 inches
Range: coasts of Europe and Asia
Habitat: tidal mudflats
Diet: small crustaceans and other invertebrates

Common seal F 4.5 inches
Range: North Atlantic, Pacific and Arctic coasts
Habitat: sheltered coastal waters
Diet: fish, squid, crustaceans

Green turtle 2 inches
Range: tropical and subtropical oceans
Habitat: marine, lays eggs on sandy beaches
Diet: jellyfish, fish eggs, worm, sponges, algae, crustaceans

 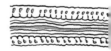

Shore crab 0.5 inches
Range: coasts worldwide
Habitat: coastal waters
Diet: mollusks, worms, crustaceans

Lugworm size varies
Range: coasts of Europe and North America
Habitat: sandy seashore
Diet: decaying creatures in sand

By freshwater

Eurasian otter F, H 2.5 inches
Range: Europe, Asia, North Africa
Habitat: lakes, streams, rivers
Diet: fish, crustaceans, amphibians, insects, birds

American beaver F 2 inches; H 6 inches
Range: North America
Habitat: streams, small lakes
Diet: wood, leaves, roots, bark

American mink F, H 1.5 inches
Range: North America, introduced to Europe
Habitat: swamps, near streams and lakes
Diet: fish, amphibians, small mammals, crayfish

Water vole F, H 1 inch
Range: Europe, Asia
Habitat: banks of streams and rivers
Diet: grasses, rushes

European common frog F 0.5 inches; H 1 inch
Range: Europe
Habitat: near water in meadows, woods, gardens
Diet: insects, snails, worms, tadpoles, algae

American crocodile F, H 8 inches
Range: North, Central and South America
Habitat: rivers and estuaries
Diet: fish, turtles, birds

INDEX

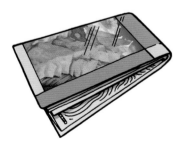

T

tarpaulin 21
teamwork 10, 36–37
tents 14
thunderstorms 26
tracking bed 155
tracking stick 153
tracings 155
trail signs 49, 114–115
trail snacks 17
tundra 28–29

V

vegetables 16
visual signaling 48–49
vitamins 16
vomiting 67, 92–93

W

water (drinking) 16–17, 24, 67
water (travel) 11, 26, 32–33
waterproof 13, 40
weather 24, 26, 28, 30
whistles 23, 131, 132–133
whole grains 16
wing prints 167
woods 22–23, 138